ACKNOWLEDGMENTS

I wish to thank the many thousands of students that I interacted with during several decades of teaching them as undergraduate or graduate students at the University of Kansas. I appreciate all that they taught me while I was teaching them.

I would like to also acknowledge with appreciation the support and assistance of my wife, Doris, who typed the manuscripts of my first books many years ago before the age of all the modern technology that authors have at their disposal today.

Understanding Severe and Unusual Weather

Joe R. Eagleman, PhD

University of Kansas

Copyright © 2021 Joe R. Eagleman

All rights reserved.

Cover Painting © 2021 Joe R. Eagleman

ISBN: 9798685910233

1

Introductory Overview

An understanding of severe and unusual weather should be a fundamental part of everyone's storehouse of knowledge. We live in a world that is at least occasionally dominated by severe and unusual weather. Many types of severe weather are sufficiently rare that a common defense mechanism of many people is to assume that they will never be directly affected. However, there is hardly a place in the whole world that does not have some peculiar aspect of weather that requires some degree of understanding and preparedness in order to avoid loss of property and, perhaps, even life itself. Fortunately, no particular location has all the different kinds of unusual and severe weather; thus, coastal areas are exposed to the tremendous power of the hurricane that bring high winds and frequently produce flood conditions, while within the interior United States, where hurricanes are not a threat, such severe types of weather as tornadoes, hailstorms, and blizzards are sufficiently frequent that an understanding of these storms is essential when traveling or living in this part of the United States. The complete destruction of a farm crop by hail is common and many houses are destroyed by tornadoes every year as shown in Figure 1-1.

Although lightning is a greater hazard in some parts of the world than others, there are very few locations, including Alaska and the Sahara Desert, where occasional severe thunderstorms do not develop numerous lightning strokes. These, of course, are a threat to

Figure 1-1 Example of houses completely destroyed by the intense winds of a tornado.

individuals who may not be properly protected, in addition to the possibility of extensive damage to property and forests, since lightning is known to be responsible for a high percentage of all forest fires. These are extremely damaging in California and Colorado as well as many other locations.

Flash flooding is an increasing hazard as flood plains become more inhabited and urban areas spread into areas prone to flooding. In recent years, flash floods have been responsible for an ever-increasing number of deaths. On the other hand, drought and prolonged lack of rainfall are common weather-related problems in various parts of the world. In many midlatitude locations unusual large-scale airflow patterns may cause prolonged drought in areas that ordinarily enjoy adequate amounts of rainfall.

There are some indications that our weather is becoming more variable with greater drought in the summer months and a return to colder winters. We are entering an age where we can begin to think of weather simulation, modification, and management, if only on a small scale at the present time. We can create and control small atmospheric vortices in the laboratory as an aid in understanding the

CONTENTS

	Acknowledgments	i
1	Introductory Overview	1
2	The Largest Storm on Earth	6
3	Blizzards and Chinooks	28
4	Setting the Stage for Severe Thunderstorms	49
5	Nature of Severe Thunderstorms	68
6	The Strongest Storm on Earth	86
7	Laboratory Tornadoes	105
8	Softening the Blow	120
9	Fire from Above	139
10	Ice from the Sky	162
11	The Mighty Middle-size Storm	181
12	Categories, Prediction and Notable Hurricanes	198
13	Floods and Drought	218
14	Unusual Storms and Weather Patterns	243
15	Human Response to Weather	259
	Index	275
	About the Author	279

Figure 1-2 The first unconfined artificial tornado created outside a confining box.

nature of weather phenomena (Figure 1-2). These are useful in studying the behavior of vortices and their relationship to various types of storms in the atmosphere. They may also furnish useful information on the nature of damaging winds within tornadoes as they destroy houses and other property.

Although we are not able to control atmospheric storms as much as we would like, there are other types of weather that we can definitely label "man-made." These include certain urban and agricultural climates. In urban areas, the climate is modified to a considerable degree by the structures that people have erected. Most of this may be called inadvertent weather modification. Changes include temperature, humidity, and various other characteristics of the environment in urban areas. Intentional weather modification has been used on a local scale in some agricultural areas in the past. The most common form of this weather modification has been cloud seeding with dry ice or silver iodide. Some have believed that certain forms of cumulus clouds can be successfully seeded to increase the amount of precipitation, and results of other research have shown that the snow-pack in mountainous areas can be augmented during the wintertime to significantly increase the amount of runoff in the spring. This activity has been used in the past but has now been largely curtailed because of environmental considerations and lack of positive results.

If we cannot directly manipulate the weather to our liking, it is still possible to protect ourselves in various ways from the damaging effects of severe and unusual weather. This can be done through proper knowledge of the various severe storms, and use of the best information on seeking shelter from them, coupled with appropriate safeguards before the severe weather arrives.

The design and construction of houses makes a great deal of difference in their ability to withstand the high winds of hurricanes or tornadoes as well as the heavy load of snow that accumulates on the roofs of many houses during a winter. This is an important protective avenue for those individuals who will someday have their own house designed and constructed. We may also be able to use the weather to our advantage by harvesting wind and sunlight. Wind power and sunshine represent two important potential sources of available energy; therefore, harvesting the wind and sunlight is an important future part of weather management in the broadest sense.

If you have never stopped to think about the magnitude of the effect of weather on us, you may be surprised to find that in this day of electronic gadgets, we are still quite frequently at the mercy of the elements. In most cases, this need not be the case if we have paid appropriate attention to the nature of the potentially severe weather events in the area where we live or travel, and have informed ourselves about the precautions that are necessary in dealing with this most severe aspect of our environment.

Thus, the blizzard can be avoided when traveling during the wintertime by realizing that some weather events must supersede our desires, and travel plans must be modified to include a delay of a day or two to allow a large snow-storm to move through the area. These frontal cyclones are generally forecast quite well by the National Weather Service in Silver Springs, Maryland.

Hurricane warnings are the responsibility of the National Hurricane Center in Miami, Florida. Information gathered there from satellites and reconnaissance flights are used for their computer models to predict which areas along the coast will be affected by the passing of a hurricane. Tornado forecasts are the responsibility of the National Severe Storms Forecast Center in Norman Oklahoma. These storms are much smaller and shorter lived than the other two vortex storms previously mentioned, and, therefore, cannot be forecast as individual tornadoes. Areas of the atmosphere that are likely to produce tornadoes, however, can be identified and such forecasts are provided by the National Severe Storms Forecast Center.

It is currently not possible to accurately predict which tropical disturbance will grow into a hurricane or which small thunderstorm will grow into a large tornado-producing thunderstorm. Thus, our knowledge of atmospheric storms and characteristics is still incomplete. We should not be discouraged by this, however, but should view this as an opportunity for future studies by ourselves or others. In this book, the nature of various severe and unusual weather phenomena will be described in detail, with emphasis placed on better understanding such weather events before they confront us, in order to provide the information needed to deal with these important, but often neglected, aspects of our environment.

2

The Largest Storm on Earth

SCALES OF ATMOSPHERIC MOTION

Imagine a beautiful snow-covered landscape, and contemplate the many different scales of magnitude that make up the scene. The entire blanket of snow including any large drifts is composed of numerous small individual snowflakes. If we reflect further, we know that an even smaller and almost separate world exists that involves the molecular arrangement of the hydrogen and oxygen atoms that form the individual snowflakes. Within the atmosphere different scales of motion, ranging from global air currents to turbulence around a building, frequently affect each other. Evidence shows that energy is transferred from the larger circulations, such as the jetstream, to smaller circulations such as a frontal cyclone that is the largest storm on earth. In a similar manner energy is transferred from a thunderstorm to a tornado where it is concentrated into a small but fantastic storm.

We will explain later the source of energy for the jetstream, but for now let's examine the source of energy for the increased wind speed in a frontal cyclone and in the tornado. First, we will look at how energy is transferred from the jetstream to smaller circulations. Let's consider a well-known physical law that governs the interaction of these atmospheric scales of motion that is called the Law of Conservation of Angular Momentum. The momentum of any moving mass is simply its velocity multiplied by its mass: for a football player his momentum is determined by how fast he is running and his mass. Momentum in an angular sense (for movement in a circular path) is related to the radius of the curved path. The Law of Conservation of Angular Momentum tells us that the velocity of a constant mass multiplied by the radius of curvature must always equal a constant value if momentum is conserved. This means that in order to maintain constant momentum, any change in the radius must result in a corresponding change in velocity. If the radius decreases the velocity must increase in a corresponding amount.

You have seen this law applied as ice skaters draw their arms in toward their body to reduce the radius of some of their mass to increase the rate of spin. It can also be demonstrated in a manner more related to atmospheric motion, with a ball on a string; in this case, the

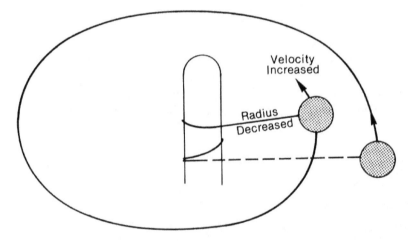

Figure 2.1 The velocity of a ball on a non-stretching string will increase as the radius of the string shortens.

radius is the length of the string (Figure 2-1). If the string is allowed to wrap around your finger as the object on the end of the string rotates, the velocity of the object increases as the string shortens.

Another physical principle (Newton's First Law of Motion) tells us that once an object acquires motion it stays in motion until forces are applied to stop it. These two very simple principles continuously operate in the atmosphere because of the various scales of motion involved.

High above the earth's surface are fast-moving streams of air that are constantly interacting with smaller scale weather phenomena in such a way that some of their momentum is transferred down to circulations having a smaller radius. As this occurs, the velocities of the smaller circulations must increase if angular momentum is conserved. This explains in a general way how the very high wind speeds experienced in smaller storms such as tornadoes are possible.

Let us look at the magnitude of some of the different scales of motion in the atmosphere. One of the largest air circulation systems is the jetstream, a portion of which may have a radius of 1000 miles. A frontal cyclone includes a rotating mass of air with a low-pressure center that receives some of its momentum from the larger air-flow patterns of the jetstream. The circulation radius of the low-pressure system may be only half that of the jetstream, or about 500 miles (Figure 2-2). Individual thunderstorms grow within a frontal cyclone and have a radius of only about 5 miles while a tornado is another order of magnitude smaller with a radius of perhaps one-half mile for even a large tornado. A few computations will convince you of the possibility of some extremely high velocities as the radii decrease if the momentum remains constant.

Using the radius of curvature just described, a frontal cyclone could contain winds of 100 mph if the jetstream were only 50 mph. If momentum is conserved and energy is fed to even smaller circulations, their calculated rotational velocities become unrealistically large.

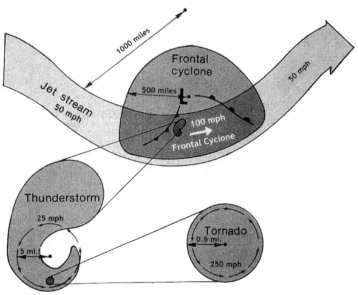

Figure 2-2. Smaller circulation in the atmosphere may acquire greater wind speeds than large circulation because of the conservation of angular momentum.

Actual speeds are much less than calculated from the jetstream to the tornado because the transfer of momentum and energy to the smaller circulations is very inefficient, and also the connection between the circulation within a frontal cyclone and a thunderstorm is very weak.

The actual mechanics of airflow around thunderstorms are to be discussed in a later chapter, but for now a realistic calculation can be made at this point for the tornado by starting with a rotational speed of a thunderstorm even less than the speed of the jetstream flowing around it. If the radius of rotation of the thunderstorm is 5 mi. and the circulation speed is only half that of the jetstream, 25 mph, for example, the speed of the tornadic winds could be 250 mph for a very large tornado and even higher for smaller tornadoes if momentum were conserved.

JETSTREAM

The larger atmospheric circulations must also have an energy source. This source is the sun, as it provides the energy for weather systems. The more direct radiation from the sun over equatorial regions is absorbed there in large amounts to cause tremendous heating, while polar regions experience low energy levels due to lack of sunshine. This creates a large difference in the amount of heat in different parts of the atmosphere. In fact, satellite measurements for a five-year period show excessive heating of the entire area from 32° N latitude to the Equator while net cooling occurs from this latitude to the North Pole (Figure 2-3). Similar differences develop in the Southern Hemisphere.

The amount of energy involved in the unequal absorption of sunlight is tremendous. Heat must be continuously transported northward from equatorial regions or they would continually get hotter, while polar regions would continue to cool. The midlatitudes provide the locus for energy transport; hence the higher incidence of storms and active weather systems in this area. Atmospheric energy differences are equalized by heat transported in hot winds moving northward, by cold air moving southward, and through heat associated with water vapor. Energy is absorbed as water evaporates and is released as water condenses to liquid or to the solid form within clouds. This energy (latent heat) amounts to about 575 calories for every gram of raindrops condensed and 677 calories for every gram of snowflakes that form. The amount of heat transferred across midlatitudes averages an astronomical 10^{20} calories per day. The actual transfer of heat is not

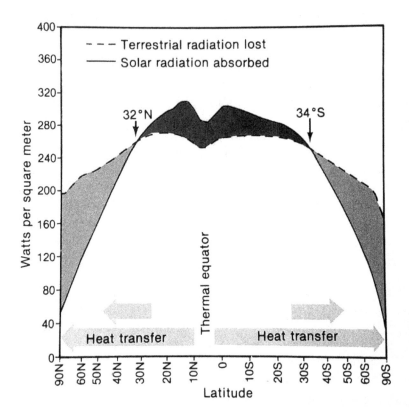

Figure 2-3. Comparison of the amount of radiation received and lost at various latitudes reveals that net cooling occurs poleward from 32° latitude. Heat is transported from the tropics to maintain an equilibrium.

constant each day as periodic weather disturbances are involved. The transfer of heat is made possible through wind currents since they respond to temperature differences.

Winds in the atmosphere blow because of differences in atmospheric pressure. In tropical regions, with a large heat input, the air expands and becomes less dense; in polar areas where heating is decreased, the air is compressed and very dense. The expanding air in the tropics rises and starts to flow northward over the northern hemisphere and southward over the southern hemisphere. In the northern hemisphere, the flow is deflected to the right of its path of motion because of the Coriolis acceleration that results from the rotation of the earth. A stream of air, thus, develops and flows from west to east over midlatitudes at a height of about seven miles. This air current

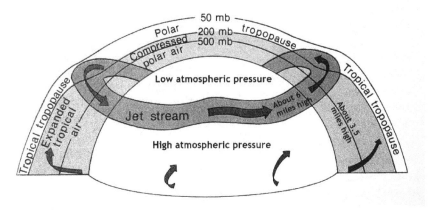

Figure 2-4 The jetstream flows about 7 miles above the earth because the greatest pressure gradient between the warm expanded air to the south and the cold air to the north exists there.

is called the **jetstream**. It is related to the surplus heating in tropical regions and the heat deficit in polar regions. This stream of air serves as a powerful source of energy which is a derivative of the primary source, solar energy.

Detailed measurements of the jetstream show that its location is related to the boundary between the expanded tropical and the denser polar air. If we consider the nature of a vertical slice of air extending from the Equator to the North Pole, the heated air near the Equator is expanded with greater vertical distance between constant pressure levels. The jetstream flows where the horizontal pressure gradient is the greatest and this occurs above midlatitudes about seven miles above the ground. The polar jet which separates these two contrasting air masses affects all midlatitude weather systems. In fact, many surface weather systems are generated by the appropriate dynamics of the jetstream that will be described in later sections.

The jetstream is wide enough to reach from Topeka, Kansas, to Oklahoma City, Oklahoma, for example, as shown in Figure 2-5, a distance of about 300 miles. The jetstream may be considered to have shells of decreasing wind velocity surrounding the inner core where the greatest wind speed is located. Outward from the inner core the wind speed decreases gradually. The velocities shown are not high since the greatest recorded jetstream speeds are about 300 mph. The average speed is about 40 mph in the summer and 80 mph in the winter. Its speed is related to solar heating. In winter, the North Pole is tilted away from the sun and the contrast in air temperature is

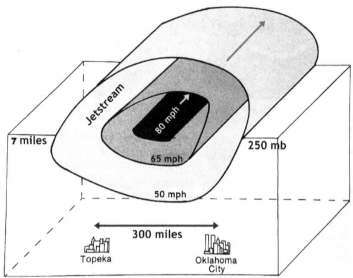

Figure 2-5. The jetstream has a central core that contains wind speeds of perhaps 80 mph. This core wind speed varies with time and location. Surrounding the central core are winds of lesser speeds.

much greater, thus strengthening the jetstream.

The position of the jetstream changes with season. During the summer, it moves farther northward, and then comes back southward during the winter. Midlatitude cyclone activity and tornadoes migrate with the season because of this phenomenon. If these storms occur during the winter, they are likely to be located farther south than during the summer.

The jetstream does not always travel the same path even during a particular season. It frequently goes through cycles ranging from west-east flow to a meandering pattern that becomes more pronounced with time until the bends eventually are bypassed by the airflow, thus, returning the jetstream to a more west-east flow. It may take the jetstream a month to go through such a cycle or it may become stationary in one of these patterns for the duration of a season. If during the summer, the jetstream develops a persistent ridge (a region of anticyclonic or clockwise curvature) over the United States this causes drought conditions.

On the other hand, a trough in the jetstream creates very wet weather. The trough is a region of the jetstream with cyclonic or counterclockwise curvature. If a trough becomes located over an area, and is stationary, midlatitude cyclones and repeated cloudy skies with more precipitation

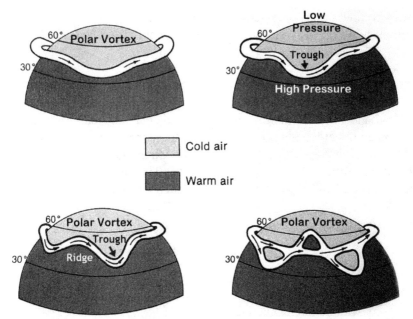

Figure 2-6. Cycles of the jetstream are repeated where it goes from straight to large meanders with large masses of air being exchanged northward or southward.

are generated. The warm air tends to remain to the south of the jetstream with cold air to the north since this arrangement maintains the strength of the jetstream. As the jetstream goes through cycles, large masses of air are redistributed as meanders in the jetstream allow cold air to move southward while the warm air moves northward as ridges in the jetstream push northward.

The jetstream is the southern edge of the **polar vortex** that exists because of the lower pressure at the poles compared to the warmer more expanded air at other latitudes.

As large meanders develop in later stages of the jetstream as shown in Figure 2-6, unusual weather occurs in various locations. For example, Alaska may be experiencing quite warm weather while the central United States has very cold weather because of large meanders in the jetstream. Very cold weather is typically associated with a large trough in the jetstream that brings extremely cold air southward. The jetstream is related to the air masses in a dual fashion since differences in air temperature contribute to the development of the jetstream, but after the jetstream gains momentum, with winds of perhaps 100 mph, it is influential in redistributing heat through the movement of large pools of air, as well as in generating storms. Thus, differences in temperature are required to generate the jetstream, but it also exerts an important influence on the movement of warm and cold air after it develops.

AIR MASSES

Air masses are simply large pools of uniform air, usually occupying a thousand miles or more. Since a particular air mass has similar temperature and humidity throughout, these properties are used to categorize air masses. A continental (c) air mass is very dry while a maritime (m) air mass is quite humid. This humidity characteristic was acquired over the source region, such as over the ocean where

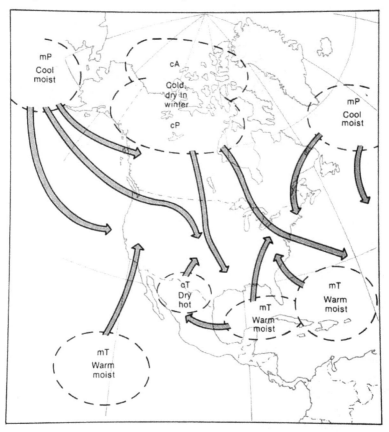

Figure 2-7. Air masses acquire their particular temperature and humidity characteristics from within their source regions as shown here. As they move out of their source regions, they affect the weather of large areas.

considerable evaporation of water increases the humidity of the maritime air or over land where the air mass acquires much less humidity. Temperature characteristics are also acquired in an air mass's source region. A continental air mass can be either tropical (T) or polar (P), depending on its source region that determines its hot or cold temperature. Maritime air masses are also either tropical (hot) or polar (cold). The four major designations of air masses, therefore, are: mT, cT, mP, and cP. Additional air mass types are occasionally important in midlatitudes including continental Arctic (A) from north of the Arctic Circle and equatorial (E) air masses from near the equator. An invasion of Arctic air causes temperatures to fall to very low values in the winter. The temperature and humidity of the various air masses are quite different in the upper troposphere as well as near the earth's surface. Air from Miami, for example, is quite hot at the surface and decreases in temperature above the surface while continental polar air masses from Canada are much colder at the surface with corresponding colder temperatures in the upper atmosphere.

The major source regions (Figure 2-7) for the United States are the Gulf of Mexico for maritime tropical air masses, from which most of the precipitation east of the Rockies originates; the Desert Southwest for summertime continental tropical air, which is quite warm, with about the same temperature as the Gulf air, but with much less humidity; the Pacific for maritime tropical air, but the mountainous regions limit its influence to mainly along the west coast; the Pacific Ocean for maritime polar air, which sometimes crosses Canada and moves across the Eastern United States; Canada for continental polar air masses; and northern Canada and the Arctic for the dryer, and extremely cold, Arctic air that occasionally reaches the United States. Because of the prevailing wind directions from the west in the upper atmosphere over the United States, the influence of maritime tropical air from the Atlantic does not extend far inland. But this air mass source is very important along the East Coast as low pressure systems move northward and bring it inland.

Air masses are very significant in determining the weather for a particular location and are determined in real time from satellites measurements. Air masses for a particular day are shown in Figure 2-8 for all of North and South America.

FRONTS

As you are well aware, from watching the TV weatherman, fronts generally usher in a different type of weather. Weather fronts represent

16 Jan 2021 20:00Z NESDIS/STAR GOES-East AirMass

Figure 2-8 Air masses for all of North and South America are tracked in real time as shown here for January 16, 2021. (NOAA)

the boundary between warm and cold air masses; if the warm air is replacing cold air, a **warm front** exists, while a **cold front** is the leading edge of a cold air mass as it replaces warm air. The leading edge of a cold air mass takes the form of a thin wedge of cold air as it replaces a warm air mass. As a cold front advance, the trailing wedge of cold air pushes warm air upward where expansion cooling causes clouds to form (Figure 2-9). Since the warm air is advancing and flowing over cooler air as a warm front passes, the type of weather is different from that associated with a cold front. Warm fronts do not force the air to rise as rapidly, but still cause cloud formation. A warm front generates rains that may last all day, followed by warmer temperatures after the passage of the front. Such warm fronts are not as frequent as cold fronts.

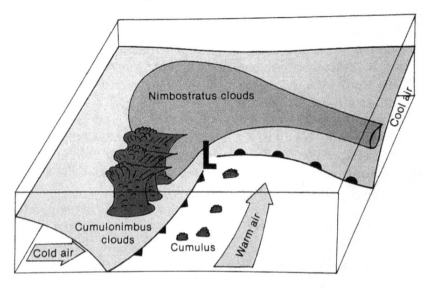

Figure 2-9. Southward moving cold air mass will create a cold front and contribute to thunderstorm formation, while a northward moving air mass will override cooler air to create a warm front and form nimbostratus clouds.

The average slope of the leading edge of cold air behind a cold front can be described in terms of its horizontal and vertical dimensions. The slope is about 1 mile vertically for every 100 miles in horizontal distance. Thus, after a cold front passes and has traveled about 100 miles past us, vertical temperature measurements would reveal warmer air about 1 mile above us. The slope of the boundary between the cold and warm air ahead of a warm front is less and causes a gentle uplift of air flowing over the wedge of cooler air. The slope is about 1 mile vertically for every 200 miles horizontally. Stratus type clouds are more likely with warm fronts, because of the gentle uplift, while thunderstorms are more likely to develop along cold fronts because of the more rapid uplift of air.

The forward speed of a typical cold front is about 20 mph; it is slower for warm fronts, about 15 mph. Either may stall and become a stationary front. In addition to temperature changes accompanying the passage of weather fronts, the wind direction also changes; normally from the southwest to the northwest after a cold front passes and from the southeast to the southwest as a warm front passes. These various types of weather fronts are typically associated with midlatitude cyclones.

MIDLATITUDE OR FRONTAL CYCLONES

A midlatitude or **frontal cyclone** consists of an area of low atmospheric pressure that forms the core of the largest type of atmospheric storm on earth. This storm contains lighter winds than smaller vortices but it is a vortex, just as a tornado is a vortex, since it contains horizontally rotating winds around a vertical core. This storm may stretch from the Canadian border to Mexico with its main influence about half that distance. A midlatitude cyclone is typically accompanied by a cold front which extends toward the southwest and a warm front which typically extends toward the east, as shown in the top map in Figure 2-10. The winds blow from the southwest within the warm air mass located to the south of the low. North of the low-pressure center winds are typically from the east, while they blow from the northwest behind the cold front. Fronts are frequently associated with the cyclone in a variety of other ways as shown in the bottom map in this figure.

A midlatitude cyclone does not develop and remain in one location, but moves across the country with a speed and direction that depends on the larger scale circulation of the atmosphere. If the jetstream is strong, the cyclone moves faster, and vice versa. The direction of movement is also related to the jetstream since frontal cyclones travel with the jetstream.

The cold front normally moves faster than the warm front and eventually catches it. This results in another type of weather front called an occluded front. As the cold front overtakes the warm front, the warm air is lifted upward and frequently continues to cause precipitation as the air swirls around the low pressure.

The center of low pressure is located by plotting the measured atmospheric pressure on a weather map, drawing lines of constant pressure, and thus delineating low pressure centers. The lowest pressure with closed lines of constant pressure corresponds to the center of a midlatitude cyclone. Cold fronts may be located by looking for regions where the wind directions shift from the southwest to the northwest. A line separating such wind directions signifies a cold front although the temperature field must also be inspected. The temperature should be warmer southward from a warm front and colder northward or westward from a cold front.

The movement of a midlatitude cyclone can be forecast in several ways. One method is simply persistence; consider the past speed and direction of the cyclone and then project this trend into the future.

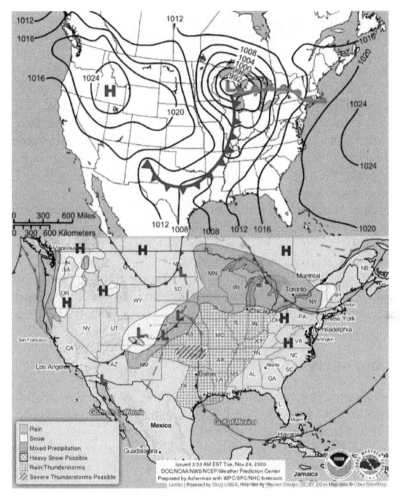

Figure 2-10. A typical frontal cyclone (top map) has a warm front extending eastward and a cold front extending southwestward. Fronts are frequently associated with cyclones in various other ways, however, as shown in the bottom map.

Another method consists of using the winds in the upper atmosphere just below the jet stream level. Average surface atmospheric pressure is 1013 mb. Millibars (mb) are used in meteorology for pressure as opposed to pounds per square inch. Since the surface pressure is approximately 1000 mb the 500 mb level is halfway through the atmosphere and is located at an altitude of about 18,000 feet. The flow of air at this height is similar in direction, but slower in speed than the flow of air at the jetstream that is at the 250 mb level about 7 miles up. Airflow at 500 mb can be used to forecast the speed of a cyclone located beneath a trough in the jetstream by projecting the cyclone forward at 50% of this speed. For example, if the air is moving at 60 mph at 500 mb, then the movement of the low would be projected at 30mph, because the surface movement tends to be about half the wind speed at 500 mb. The National Weather Service, of course, forecasts storm movement by much more sophisticated computer methods.

If we consider the airflow in a midlatitude cyclone, the surface air flows slightly toward the center of lower pressure as it circulates around it cyclonically or counterclockwise. The surface inflow toward the center feeds the large updraft located over the low-pressure area. This updraft is caused by an outflow of air (such as diverging air currents) in the upper atmosphere which then results in an updraft beneath it with inflow of air at the surface to feed it. The ascending air within the large updraft area is cooled by expansion into lower pressure at higher altitudes with resulting cloud formation.

A high-pressure area (anticyclone) has just the opposite airflow patterns with converging air in the upper atmosphere that results in descending air above the high-pressure center and outflow of diverging air at the surface.

CYCLOGENESIS

Midlatitude cyclones are produced by the appropriate conditions with the formation process called **cyclogenesis.** Cyclogenesis occurs when the jetstream provides the right conditions for producing lower pressure at the surface. This lower pressure causes the surrounding surface air to flow in toward the lower pressure and this is the beginning of a large vortex.

A number of factors are operating in the upper atmosphere to initiate cyclogenesis; an important one is upper air divergence (Figure 2-11). Diverging air results from air currents with paths that separate farther and farther apart. The extreme case is air currents traveling in completely opposite directions away from each other; normally, this does not occur over very large areas. However, air currents frequently take slightly different non-parallel paths that result in an outflow of air in those regions where the air currents are diverging. Thus, wind currents that are spreading apart produce directional divergence and this results in less air above the earth; hence, lower pressure unless compensating converging air currents are present beneath the upper air

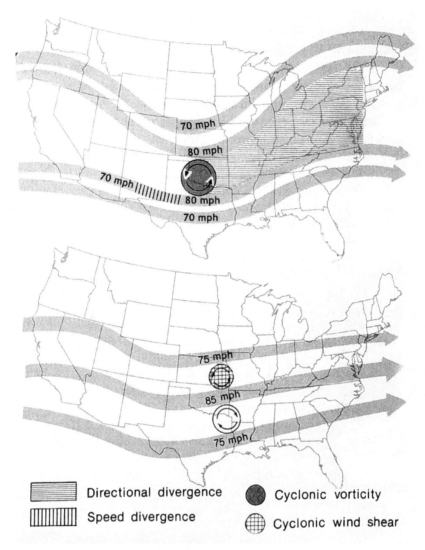

Figure 2-11. Upper airflow dynamics that initiate cyclogenesis consist of directional and speed divergence, cyclonic vorticity, and wind shear. These occur in specific regions of the jetstream as shown here.

divergence.

Speed divergence produces the same result where differences in speed exist. If, for example, a 70 mph wind is blowing over one location while the wind speed downstream is 80 mph, outflow of air occurs between these locations if the wind currents are traveling in the same direction. Thus, speed divergence affects those regions located beneath upper air streams where the wind speed increases. If we consider typical flow patterns from an upper air ridge downwind to a trough, converging air currents are typical because the streamlines corresponding to the paths of air currents are more tightly packed together in the trough than near the ridge due to greater pressure changes over shorter distances outward from a low. Thus, directional convergence occurs from the ridge to the trough, with directional divergence from the trough downwind to the next ridge.

Wind speeds are frequently greater in the trough than in the ridge which causes some speed convergence from the trough to the ridge and speed divergence from the ridge to the trough. These opposing factors may counterbalance each other, but more frequently the directional divergence is greater than the speed convergence. Therefore, regions from the trough to an upper air ridge frequently contain divergence that results in lower surface pressure, since less air is located above that point as outflow occurs. A drop in surface pressure produces winds around it because of the pressure difference, and cyclogenesis has occurred.

The jetstream exerts another influence on cyclogenesis by providing cyclonic circulation, also called cyclonic **vorticity**. Vorticity within the atmosphere can be visualized by considering a large disk placed horizontally in the atmospheric flow patterns as shown in the top map by the dark shaded circle in Figure 2-11. If it were located in a trough within a flow field of 80 mph, for example, it would begin to rotate cyclonically because the curved wind currents would exert more force on the southern side than on the northern side of the disk due to the greater length of contact as shown in this figure. Such vorticity tends to initiate smaller scale cyclonic circulation. Thus, jetstreams with their large-scale cyclonic curvature tend to induce smaller scale cyclonic circulation within the trough of the jetstream.

The jetstream frequently exerts still another influence on cyclogenesis due to wind shear in a horizontal plane which is actually another form of vorticity. If another imaginary disk were placed between two air currents flowing from west to east at 75 mph on the north side and 85 mph on the southern side, the stronger winds to the south would cause more rotation than the weaker winds on the north side, thus, producing cyclonic rotation. In this way, cyclonic circulation is initiated north of the core of the jetstream by wind shear, while anticyclonic rotation is generated south of the jetstream by wind shear.

Any one of these three primary factors, upper air divergence, cyclonic vorticity, and cyclonic wind shear, may play the major role in cyclogenesis of a particular storm or they may contribute in combination. In either case, the generation of midlatitude cyclones is determined by the jetstream. The generation region is typically in the trough of the jetstream, while their life cycle is completed in the region from the trough downwind to the next ridge as shown in Figure 2-12.

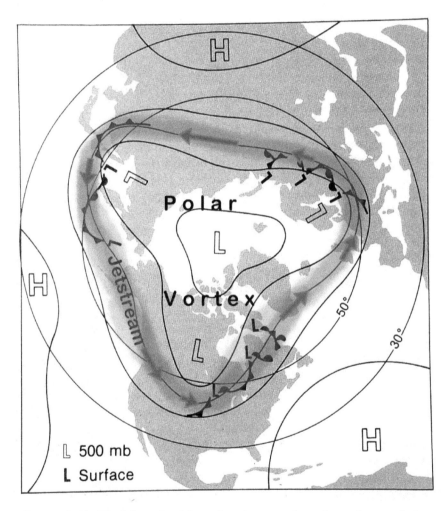

Figure 2-12. The life cycle of frontal cyclones is shown here along with the polar vortex. The polar vortex is the largest vortex on earth while the frontal cyclone is the largest storm on earth. The polar vortex is stronger in winter.

ASSOCIATED WEATHER

Frontal cyclones bring a variety of weather with them. Some cyclones are quite weak and produce only cloudy skies with very little precipitation. Others may contain winds that approach 100 mph (Figure 2-13) and bring fog, rain, sleet, and snow. Severe thunderstorms developed within them may also produce lightning, hail, and tornadoes. Several of these associated storms will be the topic of other chapters, including blizzards to be considered in the next chapter.

Frontal cyclones influence large areas as shown in Figure 2-14. The comma shape of the cyclone is apparent from satellite photographs; the head of the comma-shaped cloud pattern corresponds to the center of low pressure, whereas the tail corresponds to the cold front. The jetstream flows over the low-pressure center or slightly south of it. Since frontal cyclones are generated near the trough of the jetstream,

Figure 2-13. Frontal cyclones may generate winds strong enough to wreck ships, as shown here. Large waves whipped by high wintry winds broke the SS ARGO Merchant in half on December 21, 1976. The 200 m, 18,700-ton tanker was bound for Salem, Massachusetts, with a cargo of 7.3 million gallons of oil when she ran aground 18 miles southeast of Nantucket Island in international waters causing a major oil spill. (Courtesy of U.S. Coast Guard.)

Figure 2-14. Satellite photographs are helpful in determining the geographical extent, characteristics and movement of frontal cyclones. They frequently have a comma shape with the head of the comma corresponding to the center of low pressure as in these three frontal cyclones on December 12, 2020. The frontal cyclone that affected the eastern United States and is now over the Atlantic has an exceptionally long tail that generally forms from a cold front. The one over the eastern United States is weak and not as well defined but a strong one is approaching the West Coast. NOAA photograph.

the earth's surface beneath the region from the trough to the ridge is characterized by wetter than normal weather. In contrast, locations beneath the region from the ridge to the trough will be exposed to drier than normal weather. This information is used by the National Weather Center for making 30-day weather outlooks, since the longwave flow patterns of the jetstream frequently remain basically unchanged for several days or weeks.

If a typical frontal cyclone is on a path that will take it just north of your location, you will experience first the southeasterly winds prior to passage of the warm front. As the warm front approaches, steady rain is likely. The rain is followed by southwesterly winds and a warmer temperature. As the cold front approaches, thunderstorms develop with brief intense downpours. After the showers pass, the air is much colder with winds from the northwest.

Weather maps are prepared regularly by the National Oceanic and Atmospheric Administration (NOAA) and the Storm Prediction Center (SPC) and can be viewed in real time on their websites.

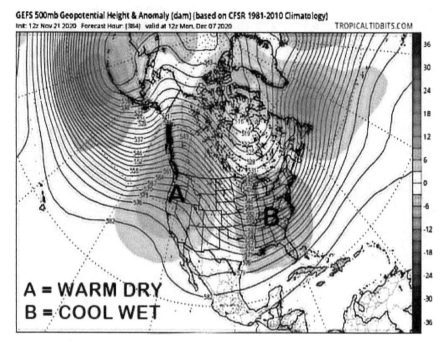

Figure 2-15 The 16 day forecast starting December 7 2020 is for a trough in the jetstream over the Eastern United States that includes wet weather and colder temperatures.

The development of troughs and ridges in the jetstream occurs over a longer time period than many other features of our weather that change on a daily basis. Thus, the long wave patterns of the jetstream are used by the National Weather service for long range forecasts. Such a forecast is shown in Figure 2-15. This 16 day forecast projects wet weather for the Eastern United States and dry, warmer weather for the Western United States. Sixty- to ninety-day forecasts are also made as shown in Figure 2-16. The pattern was forecasted to continue to some degree with the Northeastern United States getting more rain than normal over the coming three months.

SUMMARY

Many different scales of motion exist in the atmosphere. The largest vortex is the polar vortex with the jetstream at its edge. The jetstream has a large radius of curvature as it meanders around the earth

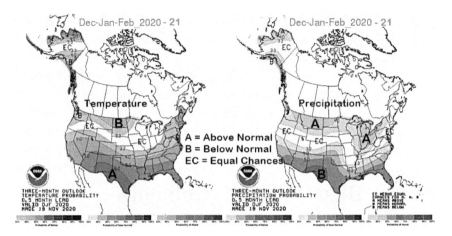

Figure 2-16 Ninety day forecast of temperature and precipitation made in November 2020 for December-February 2021.

high above the surface. It is related to differential heating of the earth by the sun, and changes location and speed with the season because of this. The conservation of angular momentum specifies that atmospheric circulations of smaller radii will have greater speeds if momentum is conserved.

Air masses having different temperature and humidity exist at different latitudes and play a significant role in weather events. Cold fronts lift air more rapidly than warm fronts and increase the chances for thunderstorms. Frontal cyclones have fronts associated with them in a variety of ways.

Cyclogenesis occurs in response to such upper air dynamics as horizontal divergence, cyclonic vorticity due to curvature of the jetstream, and cyclonic wind shear. Since these are typically associated within the region just north of a trough in the jetstream, this is the usual location of cyclogenesis.

Frontal cyclones are carried by the upper air currents. Therefore, their usual path is from the southwest toward the northeast since they are generated near cyclonic bends in the jetstream. Frontal cyclones influence large areas of the earth and frequently have the appearance of a large comma when photographed from space.

Weather maps for the surface and jetstream levels are prepared regularly by the National Oceanic and Atmospheric Administration (NOAA) and the Storm Prediction Center (SPC) where current and forecast maps can be viewed in real time on their websites.

Joe R. Eagleman

3

Blizzards and Chinooks

Although the sky was very hazy below the winter clouds, the highway was dry and the road was visible for some distance ahead. Then snowflakes began to meet the car head on, very slowly at first. But the road ahead soon vanished behind a steady stream of white dots as they dominated the air and then the ground as well. So, we will be a few minutes late, we can explain that the weather slowed us down. As the steady stream of snowflakes seemed endless, gusts of wind began to sway the car. The seriousness of the situation struck home only as the car plowed into a snowdrift, deadening the motor from snow sucked in with the air. As we surveyed the scene, it was soon apparent that it would be impossible to dig the car out of the snow, even if we were able to get the motor going again.

We had passed some houses a mile or so back; perhaps we could get to one of them to call our motor club for a tow truck since our cell phone is out of range. The first few steps were easy, but then the strong northeast wind made its presence known as it bombarded us with snowflakes and made our medium weight clothing feel very thin. As our hands and feet began to feel numb, we realized how vulnerable we had become, and began to desperately search for a way out.

BLIZZARDS

Blizzards frequently accompany severe frontal cyclones in winter, and are generally underestimated when considering severe storms, particularly those responsible for a large amount of damage and numerous deaths in the United States. Many people are surprised to learn that the number of deaths from blizzards (more than 100 per year) is similar to the number of people killed by tornadoes. A single blizzard from January 26 to 28, 1978, killed almost 100 people in the eastern United States. This blizzard was the result of a very intense low pressure system with a central surface pressure of 958 mb and winds up to 89 mph. Most of the deaths from winter storms, as well as the additional thousands of deaths each year from overexposure to ordinary cold weather, could be prevented by proper planning during winter travel and by taking other precautions.

A **blizzard warning** is issued when a frontal cyclone is expected with winds of at least 35 mph and temperatures of less than 20°F with accompanying snow. A **severe blizzard warning** is issued if the winds are expected to be at least 45 mph and temperatures are expected to drop below 10°F with snow.

TYPES OF BLIZZARDS

We described the typical frontal cyclone in the previous chapter but there are other types as well. One of the strongest that produces blizzard conditions is the **longwave cyclone** type, where the surface low pressure

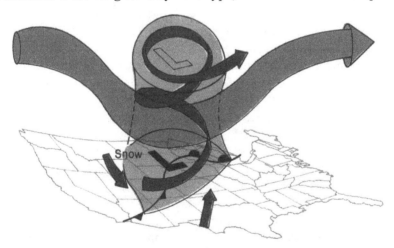

Figure 3-1. Longwave cyclones are frequently blizzard producers, since the surface winds are strong and they travel very slowly. The surface low is connected vertically with the upper-air low to form the largest of all atmospheric storms.

system is connected with the low-pressure center in the upper atmosphere instead of forming in the trough of the jetstream (Figure 3-1). The longwave cyclone develops as the jetstream flows cyclonically around the center of lower pressure in the upper atmosphere with sufficient intensity to extend this upper tropospheric vortex down to the earth's surface, thus creating very intense low pressure there. The resulting surface winds have a much greater speed as they surround the unusually low surface pressure. In addition, they are directly related to the winds of the upper troposphere since a single vortex of circulating air surrounds the common low-pressure core that extends from the surface past the jet stream.

Such longwave cyclones frequently move across the United States during the spring months, particularly March, causing much higher winds than otherwise occur with blizzard conditions in the northern United States. These storms also develop during other times of the year, and produce heavy snowfall during the coldest winter months, with high winds and cold temperatures creating blizzard conditions.

The longwave cyclone moves with a very slow forward speed because its rate of forward travel is connected directly to the rate of progressive eastward movement of the upper tropospheric longwave. Since this ordinarily is slower than the movement of other storm systems, the longwave cyclone affects a particular location for a longer period of time, and this, combined with its lower surface pressure, greater pressure gradient with stronger winds, and heavy snow, creates blizzard conditions.

An example of a very intense longwave cyclone is the storm of February 23 and 24, 1977 (Figure 3-2). On the 23rd, the low-pressure center was in northeast Kansas with winds of 25 mph common around the center; these strong winds picked up dust from the dry soil in eastern Colorado, New Mexico, Oklahoma, and Texas. The large amount of dust picked up by the high winds south of this low-pressure system was visible on satellite photographs and could be traced all the way across the southern United States and into the Atlantic Ocean. In some locations, the visibility was reduced to the point that darkness occurred in midmorning.

This midlatitude cyclone had a central pressure of 976 mb with a heavy band of snow north of the center of lower pressure. This is common for blizzards and occurs as the warm air south of the low-pressure system is carried counterclockwise around the low pressure, where it cools as it moves north of the center of low pressure, and the moisture condenses as snow. Additional precipitation in the form of

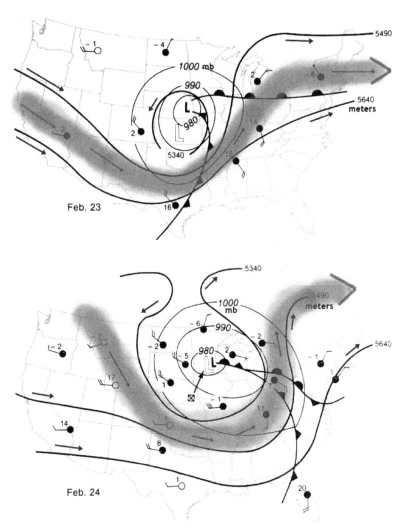

Figure 3-2. This strong longwave cyclone brought blizzard conditions to the north central United States. Winds were also so strong south of this cyclone that great quantities of dust were picked up from southwestern United States and carried across the southern United States into the Atlantic Ocean. The numbers from 980 to 1000 are surface pressure in mb and numbers like 5640 are the height of the 500 mb surface. Winds at this level are shown by arrows. Wind at the surface is shown by flags that fly opposite to wind direction.

rain may fall along the warm front, as well as along the southern parts of the cold front.

Another type of midlatitude cyclone that was described previously, the **trough cyclone**, also causes blizzard conditions, especially along the east coast. The trough cyclone (Figure 3-3) develops because of all of the upper atmospheric support features described in the previous chapter, including cyclonic vorticity, cyclonic wind shear north of the jetstream axis, and horizontal divergence of air currents in the upper troposphere. These various upper tropospheric factors contribute to an outflow of air aloft, causing a lower pressure at the surface, due to less mass of air above, and this results in air circulation around the lower pressure. These formation factors are most common in the trough of the jetstream; hence the name trough cyclone.

Trough cyclones travel forward at a higher speed than longwave

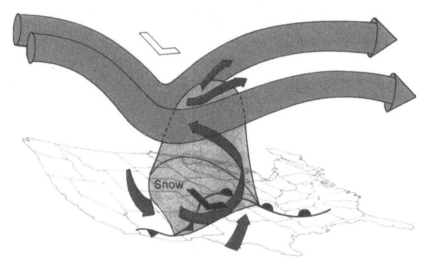

Figure 3-3. Trough cyclones develop as a result of cyclonic vorticity, cyclonic wind shear, and horizontal divergence near the jetstream. They are carried along by the upper-level winds and dissipate as they move under the influence of an upper-level ridge.

cyclones because they are carried along by the winds in the upper troposphere. In fact, their forward speed of travel can be roughly estimated by dividing the speed of the 500 mb winds by one-half. A typical speed for trough cyclones is 30 mph.

The path of a trough cyclone is also affected by the upper airstream. Trough cyclones cause blizzard conditions all along the east coast as they follow the jetstream. When a jetstream trough develops over the Gulf Coast states, it produces strong south-southwesterly flow along the east coast. This causes cyclones that develop in the trough to travel northeastward following the eastern coastline.

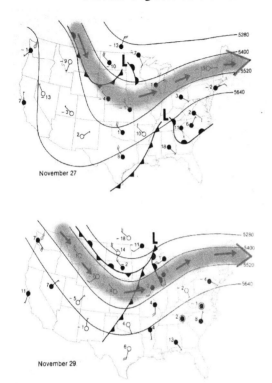

Figure 3-4. Trough cyclones that developed on November 27, 1978, were guided by the upper air stream that caused them to merge into a single, stronger cyclone two days later.

Cyclonic surface winds around the low pressure bring air from the Atlantic over the land, providing a strong moisture source that produces heavy rain and snowfall from these storms. Such **nor'easter** storms are responsible for blizzard conditions over extended areas of the heavily populated east coast.

Trough cyclones developed at the surface in two separate locations beneath the trough of the jetstream on November 27, 1978 (Figure 3-4). Their different locations beneath the upper airstreams caused their paths to merge producing a single trough cyclone two days later. This cyclone was carried along by the jetstream for another 24 hours until it reached the next ridge where it dissipated.

ASSOCIATED PRECIPITATION

Blizzards not only produce snow, but various other types of precipitation as well. The area of greatest snowfall is usually centered about 100 mi. north of the low-pressure center as moist air from the south travels cyclonically around the low-pressure center. Southward from the area of heaviest snowfall, sleet and freezing rain commonly form, while still farther southward, rain may fall. Freezing rain on the ground is possible only when the air temperature is above freezing in the cloud where the water is condensing since it must initially be in the liquid form. With surface temperatures below freezing the liquid rain

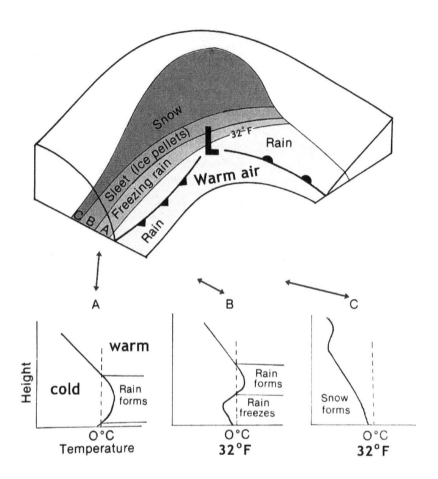

Figure 3-5. Various types of precipitation are associated with a frontal cyclone. The temperature change with height helps determine the type of precipitation at the surface. Ice pellets and freezing rain require warm air at some distance above the surface in order for the water vapor to condense in a liquid form. If the layer of air below the warm air is thick enough and cold enough to allow the rain to freeze, it becomes sleet at the surface; if not it may fall to the ground as rain, or if the surface is at or below the freezing point, it may freeze as it strikes the ground. The formation of snowflakes requires a temperature below freezing for their formation.

may quickly freeze as it strikes a cold surface. These temperature characteristics are usually associated with an inversion (increase in temperature with height) that allows the air near the surface to be cold while the air temperature is warmer at a higher level in the atmosphere.

Ice pellets (sleet) ordinarily develop from an upper atmospheric inversion. With such a temperature profile a warm layer exists, where liquid raindrops can form, above a colder layer. As rain falls down through the cold layer raindrops are frozen into ice, thus producing sleet at the surface. These appropriate temperature profiles containing surface inversions and upper-air inversions occur at varying distances behind the cold front. Surface inversions are most likely near the cold front, while upper-air inversions are likely to develop 100 miles or so behind the cold front. The average annual distributions of snow and freezing rain are shown in Figure 3-6.

A midlatitude cyclone appears as a comma-shaped storm on satellite photographs as previously described, with the cloudiness and precipitation near the center of the low pressure forming the head of the comma and the cold front producing the tail as condensation and precipitation form along the frontal system. The type of precipitation associated with a midlatitude cyclone is, of course, related to the temperature of the air. The freezing line frequently passes through the low-pressure center with the warm air south of the center and cold air to the north. This structure may change as the storm progresses and the cold air from the north flows cyclonically around the center of low pressure. Thus, blizzard conditions are most common north or west of a low-pressure center, while heavy rains fall south of the center.

WINTER STORM FORECASTING TERMS

The National Weather Service uses various terms with specific meanings in order to warn the public of the dangers associated with a winter storm. **Ice storm, freezing rain**, or **freezing drizzle** are terms used to alert the public to the possibility of a coating of ice expected to form on the ground, schools, buildings and other objects with the possibility of heavy damage. If the qualifying term, **heavy**, is used with these terms, it indicates the ice coating will be greater and more intense damage is likely to trees, wires, and other vulnerable objects.

If **snow** is forecast without a qualifying word, it means that a steady fall of snow is expected to continue for several hours. **Heavy snow** warnings are issued when a substantial accumulation is expected. This is generally 4 inches of snow or more within the next 12-hour period. **Snow flurries** are forecast when snow is expected to fall for short durations or intermittent periods. Snow flurries would not be expected to produce a very large accumulation.

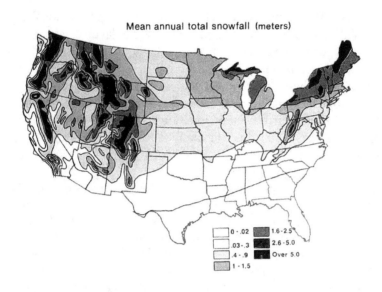

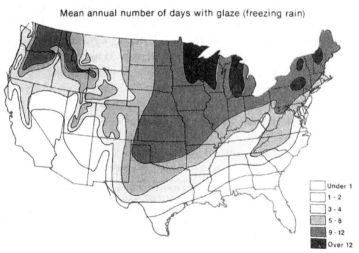

Figure 3-6. Snowfall is greatest in the mountainous areas of the western United States as orographic effects produce considerable precipitation during the winter months. The extreme southern United States gets little or no snowfall with snowfall amounts increasing northward across the central and eastern United States. The number of days with freezing rain varies from none in the extreme south and southeastern United States to an average of 12 days per year for areas around the Great Lakes and northeastern United States.

Snow squalls are brief, intense snow fall that may last for only a short period of time. They may, however, be accompanied by gusty surface winds. **Blowing and drifting snow** is forecast when strong winds accompany snow or occur when loose snow is on the ground.

Understanding Severe and Unusual Weather

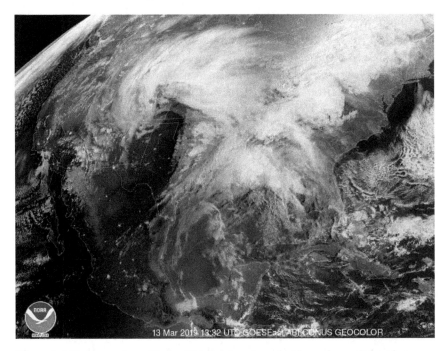

Figure 3-7 The coma shaped form of the bomb cyclone that formed over Colorado on March 13, 2019 and affected much of the United States. (NOAA Photograph)

These can reduce visibility and produce drifts that may be a problem for travel. In the Northern Plains, the combination of blowing and drifting snow after the snow has ended may be referred to as a ground blizzard.

Blizzard warnings are issued when the wind speeds are expected to be at least 35 mph with heavy snow for three or more hours. **Severe Blizzard** warnings are given with expected winds greater than 45 mph and temperatures less than $10^{\circ}F$. **Ground Blizzard** means blowing and drifting snow even if no snow is falling.

Cold Wave warnings are issued when the temperatures are expected to fall rapidly during the next 24 hours. These are issued primarily during the fall months as a warning to provide protection for agricultural crops and other commercial activities.

Bomb Cyclone warnings are issued when the pressure in the center of a frontal cyclone drops 24 mb or more in a 24-hour period. Such a reduction in pressure can result in winds of hurricane speed (greater than 74 mph).

Other terms related to the cyclone that have come into general usage are **Polar Vortex**, **Arctic Express** and **Nor'easter**. The polar vortex is the circulation around the lower pressure at the North Pole compared the higher pressure in midlatitudes as was shown in Figure 2-11. The jetstream is the circulating part of this vortex and separates the very cold air north of it from the warm air just south of the jetstream. The term Arctic Express is applied when large meanders occur in the jetstream bringing very cold air with it from its ridge to trough. Such large meanders in the jetstream can bring air from Alaska to Florida for a day or more, for example. The Nor'easter or **Northeaster** occurs when a strong frontal cyclone is located off the east coast of the United States bringing northeasterly winds from the ocean to land, north of the cyclone center. If this is in winter the humid air from the ocean brings large snowfall amounts as the air is cooled over the land. **Apparent Temperature** or **Wind Chill Temperature** is how cold the air feels due to the combined effects of wind and temperature.

EXAMPLES OF SEVERE WINTER WEATHER

A bomb cyclone formed over Colorado on March 13, 2019 as shown in Figure 3-7. This cyclone set the record for the lowest pressure ever recorded in Colorado with a central pressure of 970.4 mb at Lamar Colorado. Wind gusts greater than hurricane velocities were recorded at Denver, Colorado Springs and several other locations. Many car accidents occurred and more than 1000 people were stranded in one county alone. More than 400,000 people were without power in Colorado. This state was just one of many that was affected as the bomb cyclone made its way across the central United States. This storm resulted in damages costing over four billion dollars as it spread snow and very cold temperatures across the country.

This frontal cyclone was not the first billion-dollar cyclone. Others that caused over a billion dollars in damages were the winter storms on March 1-3, 2018, February 16-25, 2015 and January 5-8, 2014. One of the costliest was the cyclone on March 11-14 1993 with damages topping five billion dollars. One of the first billion-dollar cyclones was the winter storm of December 17-30, 1983.

The billion-dollar cyclone of January 2014 is a good example of the Arctic Express and Nor'easter. The weather map for January 6, 2014 is shown in Figure 3-8 with the location of the jetstream. The jetstream headed almost due south with winds up to 155 mph extending from North Dakota to Texas. Then, an extremely sharp bend took the air from Texas

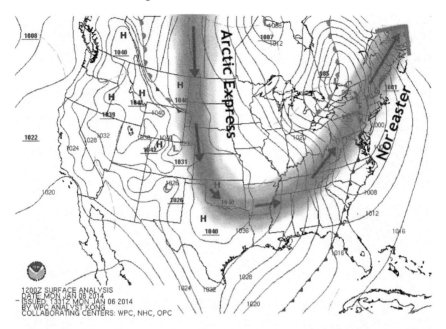

Figure 3-8 The surface weather map for January 6, 2014 with jetstream. The wind speed of the jetstream over South Dakota was 144 mph. It was over 170 mph over Maine.

to Maine with speeds up to 190 mph. Early on January 2, 2014 the system started as a low-pressure area in the Gulf of Mexico along a stationary front. Winter storm warnings began to be issued for much of the East Coast as the storm moved along the coast.

The cyclone strengthened as it moved Northeast. With its center over the ocean it became a strong Nor'easter with heavy snowfall all along the East Coast. Some of the statistics for this storm were for Boston a temperature of 2°F with a −20°F wind chill and over 7 inches of snow with 23.8 inches of snow at Boxford, Massachusetts. The Arctic Express affected Michigan, with over 11 inches of snow outside Detroit and temperatures dropping below 0°F. On January 6, 2014, Babbitt, Minnesota was the coldest place in the country at −37°F. The cold air reached as far as Dallas, which experienced a low temperature of 16°F. On January 7, 2014, the cold air reached even farther south to Houston, where the low temperature for that morning was 21°F just shy of the record-low temperature of 19°F for that day. Damages were estimated to be about five billion dollars for this storm as almost 200 million people were affected. The satellite coverage of another billion dollar Nor'easter storm on January 4, 2018 is shown in Figure 3-9.

Figure 3-9 Satellite view of the Northeaster on January 4. 2018. (NOAA Photograph)

Stationary fronts are frequently thought of in connection with flooding, but they can also produce heavy ice. This occurred on March 4, 1976, throughout the United States in a region extending from Colorado to New York. This is still referred to as "The Great Ice Storm of 1976." A trough in the upper troposphere centered over Arizona produced southwesterly winds extending from Texas through Pennsylvania (Figure 3-10). In typical fashion, the stationary front also extended from Texas through Pennsylvania since it was located parallel to the jetstream and, therefore, had no force to move it. The front remained stationary, setting up the situation for continued icy conditions. Precipitation occurred as southerly winds south of the stationary front brought warm moist air over the frontal surface to be opposed by northerly surface winds. This caused lifting of the moist

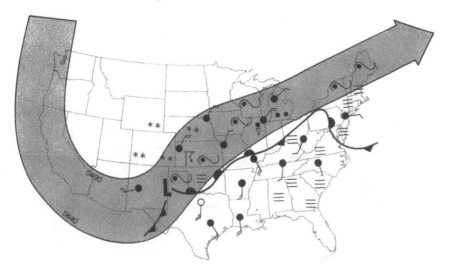

Figure 3-10. Freezing rain fell on March 4, 1976, over an area stretching from Texas to Maine as a front stalled in a parallel position to the jetstream. Continued south winds southward from the front produced a prolonged ice storm.

air, condensation, and severe problems from thunderstorms, icing, and snowfall.

The Great Ice Storm knocked television stations off the air, caused Madison reservoirs to dry up and left more than 600,000 Wisconsin residents without power for days. It started as rain on Monday, March 1, and kept up all week. Temperatures near freezing were too cold to melt the ice. A Wisconsin reporter stated in 2019 "Massive snow storms aren't exactly that rare here in Wisconsin. We're quite used to racking up a couple of feet of snow at least once a year somewhere in the state. Sure, sometimes those storms come with gusting winds or drifting snow, but honestly, a blizzard in Wisconsin is fairly commonplace at this point. But in March 1976, a devastating ice storm hit that affected nearly the entire state. It devastated counties, left many without power and is still remembered as one of the worst natural disasters in our state's history. Snow is one thing, but ice is its own beast, which is why this storm is remembered by many as the largest blizzard in Wisconsin."

Lower Michigan was also ravaged by ice five inches thick in some locations. The power in many locations was out for weeks. In New York this storm is still listed as one of the costliest ice storms in history with ice over four inches thick in some locations. Several counties in New York

Figure 3-11 The Great Ice Storm broke power lines, trees and causes many automobile accidents in addition to the many other problems it brought. (NOAA Photograph)

were declared major disaster areas by President Ford. The storm left more than 100,000 families in New York without power and at least 10,000 without water.

SURVIVAL IN BLIZZARDS

Hypothermia is the name given to the rapid loss of body heat. A person may survive for days, or even weeks, without food and with very little water, but he cannot survive for very long without body heat. Heat loss is also dependent on other factors in addition to air temperatures. The combination of high winds and cold temperatures can cause very rapid loss of heat. It is possible for a person with wet clothing to freeze to death at temperatures of 50°F. Water conducts heat away very rapidly, and provides additional cooling as latent heat is extracted from skin and clothing for evaporation.

The National Weather Service calculates the apparent temperature based on measured temperature and wind speed. It is commonly reported in the form of wind chill or apparent temperature. A temperature of 30°F and a wind velocity of 20 mph has the same effect (apparent temperature) as a temperature of 20°F and a wind velocity of only 10 mph. Exposed legs, face, or hands will freeze at temperatures of -10°F. If the temperature is below -30°F considerable danger of frostbite exists, and this danger increases as the wind speed increases. Such low temperatures will cause flesh covered by ordinary clothing to freeze in only 60 seconds if the wind speeds are very high. A person exposed to very cold temperatures may experience painful tingling, then numbing of the limbs, as their temperature drops below normal body temperature. With continued cooling this will be followed by hallucinations, drowsiness, and death.

Each year thousands of elderly Americans, along with motorists, hikers, and other outdoor enthusiasts, succumb to overexposure to the cold. The chances for accidental hypothermia are greatly increased by several other factors, such as inadequate clothing, inactivity, loss of sleep, little body fat, and the consumption of alcohol, barbiturates, and tranquilizers.

A person's limbs may be able to withstand a drop in temperature of 30°F, but if the internal body temperature drops more than about 10°F, death will result. Body heat is conserved as surface blood vessels are constricted by normal response mechanisms. The increased danger of hypothermia with the use of alcohol or certain other drugs results from their opposite action on blood vessels by expansion and increased loss of heat.

Frostbite occurs as the fluid between body cells freezes and the cells become dehydrated. This process is accompanied by feelings of extreme cold, burning sensations, and then numbing. For centuries the suggested treatment has been rubbing the frostbitten area with snow or ice. More recent tests have shown, however, that it is much better to rewarm the frozen tissue rapidly, in warm water or by another warm part of the body. Frozen tissue should not be rubbed as it can be damaged easily. Overheating a frozen part should, likewise, be avoided.

Prevention of frostbite is far better than suffering the potential consequences that include stunted bone growth or amputation of a limb, besides the possibility of death from hypothermia. Several layers of clothing are more efficient in sealing the body from heat loss than single heavy garments. While gloves are better for driving and other activities, mittens are much better for protecting fingers from frostbite while outdoors. Several pairs of socks are also helpful in protecting the feet from frostbite.

Figure 3-12. Driving can become hazardous or impossible during winter snow storms. It is sometimes better to delay travel plans than to risk the consequences of travel during blizzard conditions.

Traveling in severe storms can be extremely serious business as is evident in Figure 3-12. It is advisable to plan winter trips accordingly, realizing that blizzards are one of the greatest potential killers of people, as well as livestock and wildlife. Rational winter travel demands that the trip be planned in advance with possible alternate routes. When travel is risky you should estimate the time, it will take for the trip and give information to someone else concerning your expected arrival time. It is very important to consult weather information and to have a vehicle that is properly equipped. If a blizzard is forecast, with traveler's warnings issued, and a trip is absolutely necessary, then it is advisable to take someone else along or travel with another vehicle if possible. The increased cell-phone reception coverage in today's world has decreased the risk of travel during winter storms but it has not eliminated it.

Driving on snow or ice requires different techniques, such as starting, steering, and braking very gently. You should get the feel of the road and allow plenty of room for stopping. A gentle pumping of the brakes is usually more effective than hard braking that may cause locking of the wheels with skidding. Good visibility is important with winter driving; windshields and rear windows should be clear so that your visibility is not reduced to a peephole through the glass. Headlights should be used while traveling rather than parking lights. Snow tires and four-wheel drive are helpful in driving on snow or ice. A point to remember is that ice is much slicker near the freezing temperature than with much lower temperatures. A braking distance of twice as long is required on glazed ice at 30°F than at 0°F.

If you become stuck in snow, do not panic, and beware of overexertion through such efforts as rapid shoveling or trying to manually push the car out of a snowdrift. It is usually best to stay in the vehicle unless you are very close to help that is visible from your automobile. Disorientation and extreme fatigue can occur quickly in blowing and drifting snow; with very cold temperatures this can be disastrous. Remember that carbon monoxide poisoning can be fatal; therefore, it is advisable to allow fresh air into a car at intervals while running the motor and heater sparingly. This is also necessary because a full tank of gas will last only a few hours if run continuously. It is advisable to leave the dome lights or emergency lights on so that the car can be seen by others that may be approaching on the road.

CHINOOKS

If you become completely stranded in a blizzard, you can hope for rescue or for the development of a chinook wind. A **chinook** is able to remove a snow cover quite rapidly, because it can suddenly increase the air temperature to above freezing but they only form in particular locations. Chinook winds form under two different conditions. One of these is through the flow of air over a mountain or large topographic obstruction. As air is lifted on the windward side of a mountain, it is cooled not only by expansion of the air, but by heat added as liquid water or ice crystals condense or deposit. The net cooling rate on the windward side may be only 3°F per 1000 ft. instead of the heating rate of 5.4°F per 1000 ft. as the air descends on the leeward side of the mountain (Figure 3-13). This produces considerable heating, giving rise to a chinook wind that increases the temperature very rapidly in localized areas where the winds reach the surface. Such winds are also very low in humidity on the leeward side of the mountain.

A chinook may also develop after the air has traveled for a considerable distance over higher elevations, across the Rocky

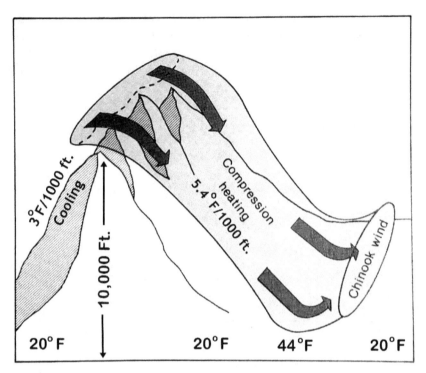

Figure 3-13. The Chinook wind may cause a rapid rise in temperature in localized areas on the leeward side of mountains as compression heating of descending air increases the air temperature and reduces the humidity.

Mountains, for example. This air may be heated by radiation with absorption of some energy as it travels across the Rocky Mountains. Then as the air suddenly descends into lower elevations in the Great Plains, compression heating at 5.4°F for every thousand feet that it descends, results in a chinook wind.

Some examples of the development of chinook winds include very rapid temperature increases at Rapid City and Spearfish, South Dakota. In January, 1911, the temperature rose from 14°F to 43°F, a temperature change of 29° in only ten minutes. On January 22, 1943, the temperature changed from -4°F to 45°F with a change of 49°F in only 2 minutes; a change from Arctic air to snow melting temperature.

Such rapid changes in temperature have been thought in the past to produce psychological effects. In fact, a similar wind that occurs in Germany and in the Alps, called a foehn wind, has been blamed for such great psychological effects that persons committing robberies or other crimes during the foehn wind have not been charged with criminal offenses. Some doctors have suggested that a change in the concentration of positive and negative ions of the air occurs during the chinook wind; this then could be responsible for people reacting in abnormal ways.

PREPARATION FOR WINTER STORMS

Since chinook winds are very localized wind systems, it is much more advisable to plan for dealing with winter storms in some other way. A winter storm car kit is advised for the wise motorist who is anticipating cross-country travel or other travel during a blizzard forecast. A winter storm car kit should include such items as blankets, sleeping bags, or even boxes of newspapers, since they have considerable insulating effect. Extra clothing such as caps, mittens, and overshoes may come in handy, as may high calorie, nonperishable food. Small cans that have been previously filled with melted wax poured around a wick to form a candle, and a supply of matches could prove to be very valuable for warmth and light. Cell phones may be helpful during times of emergencies provided they are not out of range of reception. A small sack of sand, a flashlight, a first aid kit, and a shovel are also recommended for a winter storm kit. Additional items which might be quite helpful include booster cables, a fire extinguisher, and catalytic heater.

SUMMARY

Blizzards are caused by severe frontal cyclones that produce strong winds, low temperatures, and snow. The longwave cyclone moves slow because the low pressure at the earth's surface is located beneath, and is an extension of the low-pressure center in the upper atmosphere. This type of cyclone may produce severe blizzards because of the extremely low surface pressure, strong winds, and very slow rate of travel. The trough cyclone can also generate blizzard conditions as it forms beneath the trough of the jetstream and is carried along toward the ridge.

Blizzards may produce ice pellets, freezing rain, snow squalls, and blowing and drifting snow, with rain extended southward in the warmer air. Heaviest snow is typically north of the low-pressure center extending for as much as 100 miles while various other types of precipitation occur near the cold front.

Examples of severe blizzards include the bomb cyclone of 2019 that caused record low pressure and hurricane-velocity winds. Another billion dollar cyclone in January 2014 featured the Artic Express and Nor'easter. It brought a temperature of -37°F to Minnesota. The Great Ice Storm of 1976 set records and brought a glaze of ice five inches thick in some locations.

The combination of strong winds and low temperatures cause more rapid body heat loss than if winds are light. The wind chill or apparent temperature is a measure of the cooling power of the air. Traveling during severe winter storms can be dangerous without a properly equipped vehicle and proper planning. If stranded in a snow storm it is usually preferable to remain in or near the vehicle unless help is within sight.

Chinooks are a very dry wind that cause a rapid rise in air temperature on the leeward side of mountain ranges. On one occasion the temperature rose 49°F in only 2 minutes in South Dakota.

4

Setting the Stage for Severe Thunderstorms

The afternoon began with little indication that it was unusual. The hot sun beamed down with such intensity that it seemed to be extracting even more water from the parched earth to make the already humid air even more uncomfortable. As the afternoon progressed a few small cumulus clouds could be noticed in the western sky. One of these seemed to explode as it penetrated rapidly through the atmosphere. As its top began to spread into a mushroom shape, lightning punctuated the darkening sky.

The developing storm began to really show its character as the dark cloud took on a greenish cast and its base grew many small pouchy protuberances. First came the gush of cold air. Shortly afterward sheets of rain joined the blasts of air. As the sound of rain on the roof slackened, as if the storm had passed, a small thump, thump began that was hardly noticeable at first. As the thumps became louder and closer together there was no mistaking the sound of baseball-size hail that intermingled with many smaller hailstones to cover the green grass, and make the world outside unfit for man or beast. Surely the complete silence that soon soaked through the house indicated the end of the storm. But a look at the dark cloud revealed otherwise. A small dark point extending from the southern edge of the cloud showed plainly against the clear western sky. The point grew much larger and extended rapidly downward. As it touched the ground its color darkened noticeably as dark objects spiraled outward away from its base. Its slow forward movement concealed the violence contained within.

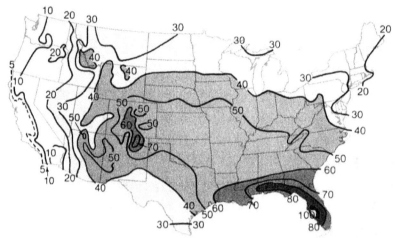

Figure 4-1. The average number of days with thunderstorms reaches a maximum in Florida where thunderstorms form on more than one hundred days out of the year. Thunderstorms are common all along the Gulf Coast. A secondary maximum of thunderstorm activity occurs in northeastern New Mexico and southern Colorado where thunderstorms form on more than seventy days out of the year. (After Environmental Data and Information Service, NOAA.)

ORDINARY AND SEVERE THUNDERSTORMS

The geographical distribution of thunderstorms is shown in Figure 4-1. They form on 100 days per year in some parts of Florida and more than 50 days per year throughout most of the central United States. The largest thunderstorms are most likely to be severe, i.e., contain frequent lightning, and damaging winds or hail. Such large thunderstorms develop in the atmosphere as localized areas of air become unstable from several factors, such as heating of the ground. As a large bubble of air becomes less dense than the surrounding air it is buoyant and rises, perhaps high enough to form a cloud. If other appropriate atmospheric conditions exist, to be described shortly, a severe thunderstorm is produced. The exact location where a severe thunderstorm will develop cannot be forecast since it is dependent upon such factors as the ground beneath it, airflow over a hill which initiates an updraft, or converging surface winds that initiate updrafts. Since the location where a specific thunderstorm will develop cannot be forecast, we use the next best approach consisting of forecasting the location where the appropriate atmospheric conditions exist for generating severe thunderstorms.

ATMOSPHERIC STABILITY

Severe thunderstorms typically occur on a day with very warm humid air, since the stability of the atmosphere is an important factor. A very unstable atmosphere develops vertical air currents easily; a stable atmosphere, on the other hand, has very few vertical air currents with no mixing in the atmosphere. Forecasters use an index to place a specific number on the stability of the atmosphere for forecasting severe thunderstorms. This index, called the lifted index, is related to the buoyancy of large bubbles of air as they rise through the surrounding atmosphere to develop a thunderstorm.

The following explanation includes the details of how the index is calculated. If you are only interested in a general explanation of severe weather forecasting then skip to the next section.

Computing the lifted index

When air changes its height and rises to a higher elevation, it cools at a specific rate called the **dry adiabatic lapse rate**. This cooling rate for unsaturated air occurs because air expands as it rises vertically because of lower air pressure at higher altitudes above the surface of the earth. A fundamental characteristic of any gas is that expansion produces cooling and compression produces heat. The expansion cooling process occurs at a constant rate as air rises because the atmospheric pressure decreases at a relatively constant rate through the troposphere. This expansion cooling rate is 5.4°F for every 1000 ft. increase in altitude. Eventually, if the rising air is cooled enough, saturation occurs and water droplets or ice crystals begin to form on particles in the air, producing a cloud.

As a cloud develops, the latent heat of condensation releases 575 calories for every gram of liquid water that condenses (679 calories for 1 gm of ice crystals). The expansion cooling rate is less as the additional heat is added. The expansion cooling rate for saturated air, therefore, consists of expansion cooling (5.4°F for every 1000 ft.) as the air rises into lower atmospheric pressure and latent heat additions as water droplets and ice crystals form in the cloud. The saturated expansion cooling rate or **moist adiabatic lapse rate** is about 3.5°F for every 1000 ft. within the typical cloud layer, but it varies somewhat for different heights.

The lifted index requires the use of the dry and moist adiabatic lapse rates in order to quantify the stability of the atmosphere for

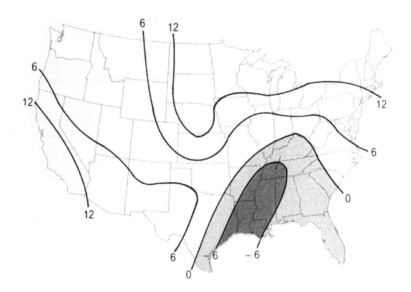

Figure 4-2. The lifted index is a number computed to represent the stability of the atmosphere. When it is negative, severe thunderstorms are more likely because of the unstable nature of the atmosphere. The lifted index was negative over the southeastern United States on April 3, 1974. More than 100 tornadoes formed out of this unstable air.

use in severe thunderstorm forecasting. The index is computed by using the average specific humidity (grams of water vapor per kilogram of air) through the layer from the surface to 1 km. Another ingredient is the maximum forecast temperature for the day since this corresponds to the most likely time of severe thunderstorm activity, through the relationship between surface heating and thermal convection. The large bubble of air having the surface forecast temperature and average specific humidity is assumed to be lifted and cooled at the dry adiabatic lapse rate until it reaches saturation at the cloud base or lifting condensation level. Further lifting proceeds at the saturated expansion cooling rate until the 500 mb level (about 18000 ft.) is reached.

The temperature of the theoretically lifted blob of air is compared with the measured temperature at the 500 mb level to obtain the lifted index. The measured temperature is available from weather balloons carrying radiosondes that are sent up twice a day from numerous weather stations in the United States to measure the temperature, humidity and pressure at various heights, including the 500 mb level. Thus, measurements are available for determining the lifted index by subtracting the temperature of the lifted blob of air from the measured temperature of the atmosphere at 500 mb. If the lifted index is negative, it means the temperature inside the blob of air is greater than the surrounding atmosphere; thus, the atmosphere is more unstable with severe weather likely. A positive lifted index, on the other hand, means that the atmosphere is stable and the growth of thunderstorms is unlikely.

During the time of the devastating Topeka tornado on June 8, 1966, the lifted index was -6 at Topeka and -4 throughout eastern Kansas and central Oklahoma as two tornadoes formed, one in Topeka and one in central Oklahoma. At 6:00 AM on April 3, 1974, on the day of the worst tornado activity since records have been kept most of the southeastern United States had an unstable atmosphere (Figure 4-2). The unstable region moved northward during the day as 148 tornadoes formed in 11 states east of the Mississippi river.

The lifted index has been used for many years for thunderstorm forecasting, however, another stability index the Convective Available Potential Energy (CAPE) is considered by most as a superior measurement of instability and is preferred by many meteorologists for convection forecasting. However, the lifted index is easier and faster to determine without using a computer, as determining CAPE requires integration from one level to another.

LOW-LEVEL JET

The nocturnal low-level jet is a wind that is characterized by a strong, near surface southerly wind that transports warm, moist air from the Gulf of Mexico northward. This moisture and temperature advection promotes instability and enhanced low-level convergence for vertical motion, promoting the development of thunderstorms. A strong diurnal oscillation occurs with the strongest wind speeds at night. The average height of the low-level jet is about 2000 ft. This is near the level of the nocturnal inversion (layer with increase in temperature with height) that is also a factor in severe thunderstorm growth and will be explained later.

A low-level jet commonly flows northward across the Great Plains at a much lower level than the polar jetstream, which flows from west to east at about 40,000 ft. as previously described. The low-level jet is considered a jetstream only in the sense that the airstream has a higher speed than the surrounding air. The low-level jet is related to lower pressure northward with the air flowing around and slightly toward

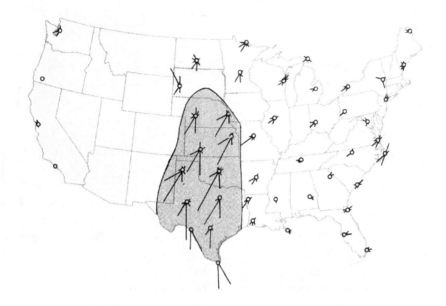

Figure 4-3. Observations of the low level jet have shown that it occurs more frequently in some parts of the United States (shaded area). The length of the lines is related to the speed of the wind and orientation to its direction. (After Mon. Wea. Rev.)

the center of low pressure.

During the summer months the low-level jet frequently extends from Texas northward through the central states as shown in Figure 4-3. The low-level jet is important in severe thunderstorm formation, since it provides an air current from a different direction from the upper-level winds and establishes wind shear that contributes to thunderstorm development in a manner to be described in more detail in a later section. Since it is greater at night it helps explain the formation of tornadoes after dark when the heat from the sun is no longer a factor in causing the air to be more unstable.

MOISTURE TONGUE

Thunderstorms need moist air for development. As a general rule, the surface dewpoint temperature needs to be 55°F or greater for thunderstorms to develop. A vast source of very moist air exists over the Gulf of Mexico that streams northward into the Central United States when the wind currents are appropriate. This happens when a low-pressure area moves into the Central United States, with stronger

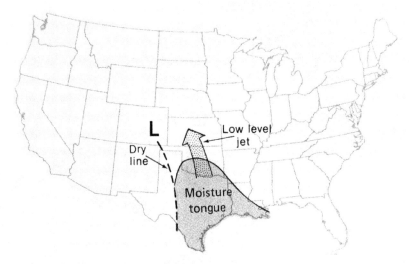

Figure 4-4. The moisture tongue, dry line, and low-level jet are shown here near the time of formation of the very devastating Wichita Falls, Texas, tornado. The western boundary of the moisture tongue forms a dry line that separates the Maritime Tropical air from the very dry Continental Tropical air.

south winds on the southeastern side of the low as air flows cyclonically around it. This reinforces the low-level jet and brings the moist Gulf air northward as a tongue-shaped stream of moist air (Figure 4-4). A sharp boundary frequently exists between the Maritime Tropical air from the Gulf and the Continental Tropical air from the Southwest. This boundary is called a **dry line** and is the western boundary of the moisture tongue. Since the drier air from the Southwest is denser than the moist air (molecular weight of water vapor is only 18, while the average of the gases that form dry air is 28.9) with a similar temperature, the dry line is similar to a cold front with the dry denser air analogous to the cold air. Thus, the dry line is an area where severe weather is frequently observed just as a cold front produces clouds and precipitation.

UPPER AIR INVERSION

Another factor frequently observed when severe weather develops is an upper-air inversion. Inversions are regions in the atmosphere where the temperature increases with height, in contrast to the normal decrease in temperature from the surface upward to about 40,000 ft. The temperature profile prior to severe weather formation typically shows a temperature decrease with height from the surface to about

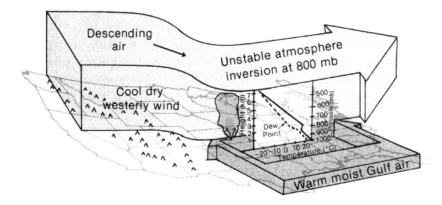

Figure 4-5. As the moist air flows northward across the central Great Plains it encounters the cool, dry, westerly winds descending over the Rocky Mountains. This creates an inversion at the boundary between the two. Small cumulus clouds may be unable to penetrate through this inversion. As a cloud eventually penetrates the inversion, it is able to grow much larger than in a normal atmosphere because of the extreme contrast in moisture and temperature properties between the two layers.

800 mb followed by a temperature increase with height for a short distance and then a temperature decrease with height. The upper air inversion layer consists of the layer where the temperature increases (Figure 4-5). The presence of an upper air inversion contributes to tornado development by suppressing cloud development until a thermal, or small cloud, is able to break through the inversion layer. This delay frequently allows the resulting thunderstorm to grow much larger than ordinary because the stable inversion layer prevents initial thunderstorm development. The delayed cloud development lets the sunlight continue to warm the ground and lower atmosphere until a bubble of air is so much warmer than the air above the inversion layer that much more violent thunderstorms form.

MIDLATITUDE CYCLONE

Severe thunderstorm and tornado watches are normally located in the warm air sector of a midlatitude cyclone. The warm air sector is the region of the cyclone south of the warm front and east of the cold front. The surface winds in this region are normally from a southwesterly direction in comparison to those behind the cold front

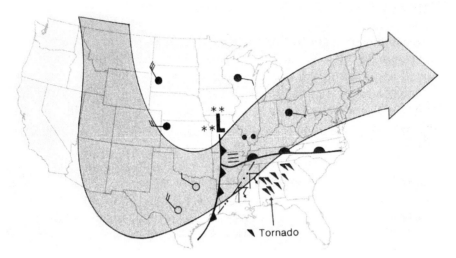

Figure 4-6. The weather map for April 4, 1977 shows the location of a frontal cyclone in the central United States. Squall lines and tornadoes developed within the warm air mass as the frontal cyclone with its associated front moved northeastward. One of the tornadoes that hit Birmingham, Alabama, caused 22 deaths and 15 million dollars' worth of damage.

from the north or northwest.

A particular type of midlatitude cyclone frequently produces severe thunderstorms. A midlatitude cyclone that is generated just north of the main axis of the jetstream, yet within the trough of the jetstream is in an ideal position to take advantage of all the various factors associated with the development of strong midlatitude cyclones. These include the cyclonic curvature of the jetstream, cyclonic wind shear just north of the jetstream (cyclonic vorticity) and horizontal divergence. Thus, the midlatitude cyclone that produces severe weather is a hybrid between the trough and longwave cyclones with the strength of a longwave cyclone and the speed of a trough cyclone. A typical severe thunderstorm producing midlatitude cyclone is shown in Figure 4-6. This particular cyclone occurred on April 4, 1977, and is situated at the surface just north of the axis of the jetstream and slightly east of the trough. Its position at this location gives it the greatest strength and allows the cold front and dry line to travel at a greater speed as they are pushed forward by the faster winds above. This speed causes more rapid uplift of the warm, moist air, and combines with other strong atmospheric characteristics to increase the likelihood of severe weather.

SQUALL LINES

Although an individual thunderstorm may become severe, the development of a squall line intensifies thunderstorm growth. A squall line is simply a line of thunderstorms, but the thunderstorms interact to reinforce each other. Squall lines develop in the warm air sector of a midlatitude cyclone; frequently along the dry line, since this is an instability zone between the warm, humid air and hot, dry air.

After a squall line has developed, we can pinpoint the most probable location of severe weather much more closely. Prior to squall line development only the large-scale atmospheric features are available to forecast the location of possible severe weather, but after the individual thunderstorms form, they can be projected to continue for several hours.

With the availability of radar apps on cell phones thunderstorm movement can be seen in real time. One of the distinguishing features of a severe thunderstorm is the hook echo on the radar screen. A good example of this is the radar image for Cincinnati shown in Figure 4-7 for April 3, 1974. This was a day with more than a hundred tornadoes. Several thunderstorms were developing to the severe stage, as they were carried from the southwest to the northeast by the upper

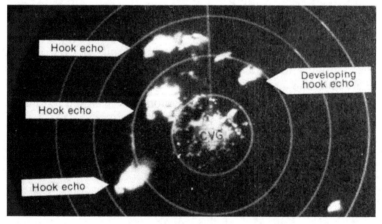

Figure 4-7. The radar screen located near Cincinnati, Ohio, revealed three severe thunderstorms with hook echoes on April 3, 1974. These thunderstorms contained very large tornadoes. The fourth thunderstorm toward the northeast of the radar site later developed the large tornado that destroyed much of Xenia, Ohio. The presence of this number of hook echoes at one time is very unusual.

Understanding Severe and Unusual Weather

airstream. Each of three thunderstorms developed three different large tornadoes about 0.5 mi. wide, and a fourth thunderstorm developed a large tornado that devastated Xenia, Ohio.

Individual thunderstorms are more likely to be severe if they are located within particular regions of a squall line. One such location is the intersection of a squall line with a cold front. The particular thunderstorm nearest this location frequently reaches the severe stage with the help of converging surface winds from the southwest ahead of the front and from the northwest behind the front. These feed air into a developing thunderstorm and intensify updrafts, thus helping the thunderstorm grow faster. Similarly, the intersection of a squall line with the dry line or the warm front is likely to produce a severe thunderstorm.

Other locations where severe thunderstorms are likely are near the central bend in a bow-shaped squall line, and at the southernmost end of a squall line, since this thunderstorm is fed by the moist air within the moisture tongue. The **Bow Echo** is known for producing strong and damaging winds. The one in Figure 4-8 brought winds of 70 mph to Chicago with corresponding damage to trees, power lines and buildings.

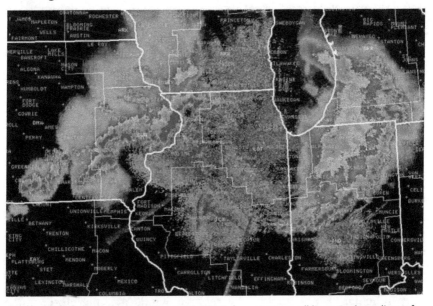

Figure 4-8. This radar image for June 18, 2010 shows two squall lines. Both are lines of thunderstorms. The one on the right is a bow echo produced from cold outflow from the thunderstorms. (NOAA Photograph)

WIND SHEAR

The proper vertical wind profile also contributes to the development of severe thunderstorms. The wind direction changes with altitude, **wind shear,** are very important since such air currents are the life streams of a thunderstorm. The winds, relative to the moving thunderstorm, are of primary importance. The concept of relative winds can be understood by considering a moving automobile. The relative wind for an automobile traveling on the highway at 70 mph, when the natural wind is calm or against the side of the automobile, is from the front of the car at 70 mph because of the movement of the car. The car creates a wind equal to its own speed in a direction opposite to its travel. This wind speed can be measured by holding an instrument outside the window.

A thunderstorm, pushed by upper-level winds, travels at a typical speed of 25 mph, producing a relative wind on the front of the storm in low-levels of this same magnitude. If the air were otherwise perfectly calm, and the thunderstorm was moving at 25 mph, the winds in the lower atmosphere would strike the front of the thunderstorm at 25 mph, creating a relative wind. The actual wind on the front of a thunderstorm is the combined relative wind and the measured wind that would hit a stationary object.

The strong winds in the upper atmosphere are responsible for the forward motion of the thunderstorm as they push it along from a southwesterly direction, under typical conditions. With a wind speed at the jetstream level of 60 mph, a large thunderstorm is not blown apart, but travels along as a unit because of air currents and rotation within the thunderstorm. With a light surface wind from the southeast and winds against the upper part of the thunderstorm at 60 mph, to produce thunderstorm movement of 25 mph, the relative wind speed at the back of the thunderstorm pushing it along would be the difference between the movement of the thunderstorm and the speed of the upper level winds from the southwest, or 35 mph. The winds at the front of the lower part of the storm would be from the northeast and almost as strong, 25 mph. The southeast wind at the surface would not change the wind at the front of the storm because it is perpendicular to movement. The relative winds are, therefore, much different from the actual measured winds above a particular location.

The proper wind profile for the development of severe thunderstorms consists of strong opposing relative winds with considerable wind shear

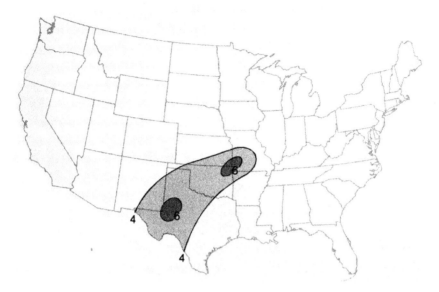

Figure 4-9. The wind shear relative to a moving thunderstorm may provide information on where tornadoes are likely to develop. A wind shear index can be computed by counting the number of 100 mb layers above 850 mb that contain winds approximately opposite in direction and equal in magnitude to the low-level winds below 850 mb. The wind shear index showed that 4 of the 6 layers oppose the low-level winds on April 11 at 00 GMT, near the time of the F4 Wichita Falls tornado in 1979. Other tornadoes developed in Texas and Oklahoma on this date.

(Figure 4-9). Wind shear exists in the atmosphere if the winds in different layers are from different directions or if a moving thunderstorm creates winds from opposing directions. Strong winds above lighter winds or winds from opposing directions cause horizontal roll clouds that may be tilted into vertical positions to become the rotating cores of severe thunderstorms. Strong wind shear is also required to support the internal thunderstorm structure to be described later. Thus, a strong wind shear environment is conducive to severe thunderstorm formation.

UPPER AIR DIVERGENCE

Atmospheric regions that contain upper air divergence are very favorable for severe thunderstorm formation. Upper air divergence occurs as airstreams spread apart and represent a region with net outflow of air (Figure 4-10). Such regions in the upper atmosphere tend to suck the air up from beneath, thus initiating or intensifying

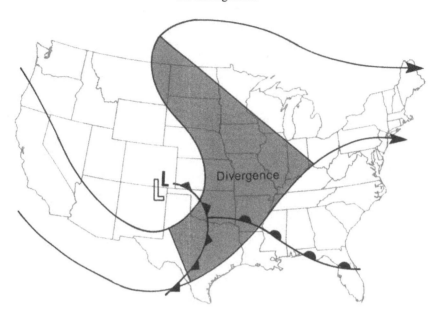

Figure 4-10. Diverging air currents above the Earth's surface are shown here for April 11, 1979 and are important in generating severe thunderstorms beneath, since updrafts located beneath areas of diverging air are likely to be much stronger. Considerable divergence existed over the central United States at the time of the Wichita Falls tornado. Tornadoes were active over six other states as well on this date with 34 of them forming from this weather pattern.

updrafts in thunderstorms located beneath such areas.

Two of the worst tornado outbreaks in history occurred on April 3, 1974, and April 11, 1965. On April 3, 148 separate tornadoes caused more than three-fourths of a billion dollars worth of damage with thousands of people injured and 265 people killed over an eleven state area. Upper air divergence was a major factor in this widespread severe thunderstorm activity. Many of the tornadoes were quite large (0.5 mi. in diameter at the base). This was very unusual but possible since all the atmospheric conditions were present for the formation of large thunderstorms. On April 11, the upper air divergence was not quite as pronounced and the trough of the jetstream was less curved than in the previous case, but considerable divergence was also present that contributed to the formation of many large thunderstorms. Thus, upper air divergence contributes to severe thunderstorm development by producing a vacuum effect in the upper atmosphere that contributes to strong updrafts.

Understanding Severe and Unusual Weather

JETSTREAM

The jetstream is a major factor in the development of severe thunderstorms. The jetstream is important because of: 1) the Conservation of Angular Momentum, discussed previously, with available energy transferred from the large circulation down to smaller circulation, thus providing a source of energy for the severe thunderstorm and tornado; 2) the influence it exerts on the midlatitude cyclone. As the jetstream goes through cycles (first as a west to east wind, then to a stream of air that meanders north and south, and eventually forms loops that are cut off before the airstream returns to a west-east flow again) it helps to generate areas of low pressure (midlatitude cyclones) near the surface (Figure 4-11). These then develop a cold front, and perhaps a dry line along with the various other atmospheric conditions that contribute to severe thunderstorm activity. On a smaller scale, the strong winds supplied by the jetstream around the thunderstorm are important in initiating and maintaining rotation within the severe thunderstorm; thereby, feeding energy directly into an individual thunderstorm.

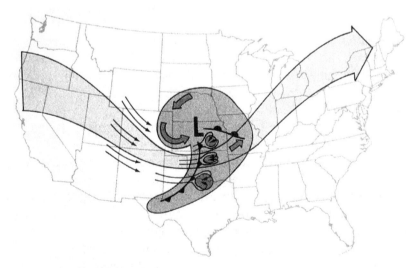

Figure 4-11. The jetstream not only helps to form the frontal cyclone within its cyclonic bend but it also feeds energy into individual thunderstorms by the strong air flow around them. Its cyclonic curvature also strengthens cyclonic vorticity (circulation) within an individual thunderstorm.

In addition to the role of the jetstream in generating midlatitude cyclones, it also determines the speed and direction of movement of the midlatitude cyclone. The typical path of a midlatitude cyclone is from the southwest to the northeast because of jetstream flow in this direction. If on a particular day a very pronounced curvature exists in the jetstream producing a northward bound jetstream, this causes a low-pressure center located beneath it to move in this same direction. If the speed of the jetstream is greater, it will move the cyclone faster and more energy will be available for smaller circulation.

SEVERE THUNDERSTORM WATCHES AND WARNINGS

A **severe thunderstorm watch** is issued by the Storm Prediction Center located in Norman, Oklahoma, on the basis of a combination of all the appropriate atmospheric conditions just described. A **tornado warning** is issued when a tornado is spotted or when a distinguishing feature such as a hook echo is seen on the radar screen. The intersection of the moisture tongue and low-level jet with a line projected downward from the axis of the jetstream is a common location of severe weather. This is typically located within the warm air sector of a midlatitude cyclone, and also corresponds to a region of upper level divergence and cyclonic curvature of the jetstream as was shown in Figure 4-11. The lifted index is frequently more negative farther southward in the warmer air. However, severe thunderstorm activity is most likely where the most divergence exists above the moisture tongue and low-level jet if an upper air inversion is present. With these conditions, squall lines develop in advance of the cold front along the dry line.

The area described in a severe thunderstorm forecast is frequently a rectangle as shown in Figure 4-12. It is located in the warm air sector of a midlatitude cyclone and may be bounded to the north by the warm front and to the west by the cold front. Since the whole weather system is moving, the line of most likely activity continually moves toward the east; therefore, a rectangle is needed to describe the forecast area of severe weather for more than an instant in time. You can frequently tell when the danger of severe thunderstorms is past before the watch is officially ended by noting the passing of the cold front. If the wind changes direction from the southwest to the northwest, and the temperature decreases, this indicates the front has

Understanding Severe and Unusual Weather

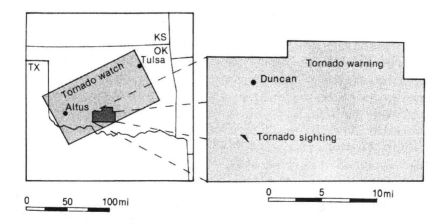

Figure 4-12. Tornado watch areas are issued on the basis of the presence of appropriate atmospheric conditions for tornado formation; thus, they cover a large geographical area. Tornado warnings cover only a few counties and are generally issued after a tornado has been sighted.

passed, with a decreased likelihood of severe weather.

Additional "inside information" is available after squall lines have developed within a severe thunderstorm watch area, since the whole watch area does not normally have thunderstorms. Existing thunderstorms can be tracked by radar and are likely to move toward the northeast because of the upper level winds. Therefore, the area most likely to have severe weather is pinpointed much more closely than the large rectangle describing the entire severe thunderstorm watch area. Once a thunderstorm develops it can be tracked on a cell phone with future movement projected from its past movement.

In some cases, a severe thunderstorm watch is issued and no storms develop. Additional triggering mechanisms such as topographic features, and uneven heating at the earth's surface are sometimes needed to start thunderstorm development.

The impact of the ordinary severe thunderstorm watch is weakened because of the size of the area. Ordinarily, 90% of the watch area may not even have cloud cover, and only one or two thunderstorms within the area develop hail, high winds, or a tornado. This makes the forecast seem much worse than it really is to the many people who do not see a cloud.

The atmospheric conditions required for severe thunderstorms vary somewhat with geographical location. In the southern United States, the atmosphere becomes so humid and unstable that severe thunderstorms can grow in the absence of some of the other specific atmospheric conditions, such as a strong jetstream, that are otherwise required for their formation. Thus, the relative magnitude of the various atmospheric conditions is important, which makes the problem much more complex than it would be if we were dealing only with the presence or absence of all the various atmospheric conditions.

TORNADOES WITHIN HURRICANES

It is not uncommon for hurricanes to spawn tornadoes, and they are similar to those that arise out of large thunderstorms described previously. When they form, tornadoes are created in the outer rain bands of hurricanes that are made up of thunderstorms. After landfall of a hurricane the very strong winds around it interact with the upper-level winds, that are coming from a different direction, creating extreme wind shear. This is one of the factors that was discussed previously in the section describing the proper wind profile for tornado development. The thunderstorms in the rain bands in front of the hurricane have such extreme wind shear between different heights in the atmosphere that this factor dominates all the other factors important for the isolated Great Plains severe thunderstorm.

The amount of information available to us in real time today on our cell phones is greater than the information available from the largest computers only a few decades ago. For example, when hurricane Delta was poised to strike the Gulf Coast states on October 9, 2020, tornado forecasts were issued for the specific region in front of the hurricane as shown in Figure 4-13. Some of the major cities included in this forecast were New Orleans Louisiana and Mobile Alabama.

SUMMARY

Severe thunderstorms generally form when several different atmospheric conditions coincide. The typical severe thunderstorm day begins with sunny skies that allow sunlight to warm the earth and lower atmosphere. A warm, moist air current flows northward over the ground from the Gulf of Mexico toward and around a low-pressure center, resulting in a low-level jet and a moisture tongue. Strong westerly flow in the upper troposphere, provided by the jetstream, brings cooler air over the Rocky Mountains. This increases the instability of the atmosphere, producing a negative lifted index and

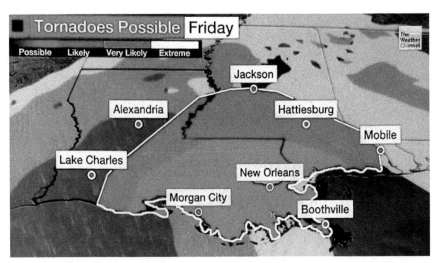

Figure 4-13. The forecast for tornadoes prior to landfall of Hurricane Delta included New Orleans and other major cities. This forecast was printed from the Weather Channel app on a cell phone on October 9, 2020 before landfall. Delta generated 14 tornadoes and caused four billion dollars damages.

creating an inversion layer.

As the midlatitude cyclone strengthens (because of its position just north of the jet stream axis in a region of cyclonic vorticity, cyclonic wind shear, and horizontal divergence) its forward movement and circulation help generate thunderstorms in advance of the cold front along the dry line. As particular thunderstorms within a squall line encounter favorable environment for growth, such as intersection with the dry line or a weather front, they grow more rapidly and become severe.

Forecasting of severe thunderstorms is the responsibility of the Storms Prediction Center as a part of NOAA. Severe thunderstorm watches are issued for any geographical region in the United States where atmospheric conditions are favorable for the growth of severe thunderstorms. Warnings are issued after a tornado has been detected on radar or has been sighted. Watches and warnings are available in real time on cell phones.

5

Nature of Severe Thunderstorms

To an avid hot-air balloonist the prospect of taking an instrument package up to the base of the large normal looking thunderstorm seemed only mildly adventuresome. The balloon rose slowly near the front of the cloud. As the balloon approached the thunderstorm a stream of air accelerated us into a notch-shaped opening. The ascension rate increased as the cloud droplets within the updraft engulfed us and began to make us feel wet and cold. As the air temperature dropped below freezing, we soon became concerned about the coating of ice that began to form everywhere. From gravity we should have been falling due to the ice formation, but our sensations told us we were still rising rapidly. Air currents seemed to be whirling in all directions around us.

The pulsating light from lightning discharges, dangerously near, revealed hailstones, carried by the whirling air currents, collecting near our feet. As the temperature dropped to 40 below zero, our concerns compounded. The extreme wind gusts were ready to tear our craft apart at any moment, and the wind and cold were beginning to chill us through our heavy clothing. At that moment the bottom seemed to fall out of the sky as we were caught by a downdraft. We knew we were much lower when liquid water droplets pelted our faces. As we emerged from the cloud, we realized that we were traveling downward at such a speed that we would be extremely lucky if we weren't dashed to bits against the ground.

THUNDERSTORM DEVELOPMENT

Thunderstorms have their origin in small cumulus clouds that begin to grow vertically through the troposphere. While cumulus clouds are quite common, only a very small percentage become large cumulonimbi and reach the severe thunderstorm stage. This stage is reached when the thunderstorm develops frequent lightning, accompanied by locally damaging winds or hail. It becomes a **severe thunderstorm** when it contains one or more of the following: hail one inch or greater, winds in excess of 58 mph, or a tornado. **Damaging winds** are defined as sustained or gusty surface winds of at least 58 mph. Severe thunderstorms also produce tornadoes, but are classified as severe thunderstorms when no hail or tornado is present if they generate locally damaging winds and frequent lightning.

The ordinary thunderstorm has a lifetime of less than one hour, because it grows through an atmosphere where the winds increase with height and the strong winds above tend to shear off its top as it builds upward. The thunderstorm that reaches the severe stage, however, must develop an internal structure that allows it to thrive on the strong upper level winds, since its lifetime is several hours and it frequently penetrates through a very strong jetstream.

When the appropriate atmospheric conditions are combined, as in a severe thunderstorm watch rectangle as described in the previous chapter, a thunderstorm may grow to the severe stage from an isolated thunderstorm, squall line thunderstorm, or multicell thunderstorm. The single isolated **supercell thunderstorm** develops as a cloud breaks through the inversion layer and grows rapidly through the stronger winds above (Figure 5-1). This frequently requires more than one trial as one cloud is eroded away by the strong winds, and a second rising turret encounters the more humid air left from the previous decayed cloud and is able to grow more rapidly through the less harsh environment.

As thunderstorms develop in a row, forming a squall line, each thunderstorm influences the adjoining one through the effect of the gust front and blockage of some of the upper level winds. The **gust front** is a miniature cold front created at the ground beneath the leading edge of a thunderstorm as a cold downdraft through the rain area spreads outward upon striking the ground as shown by the darkest area in Figure 5-1. This may reinforce the updraft of an adjoining thunderstorm by pushing warm, moist air upward. The squall line gives some degree of blockage of upper level winds. This makes it easier for new cells

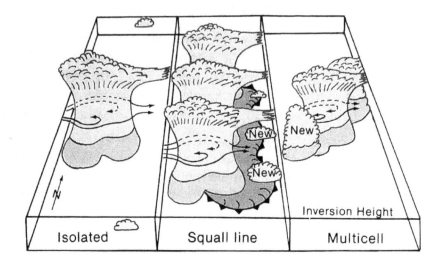

Figure 5-1. Thunderstorms may reach the severe stage as isolated thunderstorms, within squall lines, or as multicell thunderstorms.

to grow downwind from the squall line.

Multicell thunderstorms may also reach the severe stage. This type of thunderstorm grows in a cluster of perhaps three cells, where the downwind cell is decaying, the central cell is mature, and the upwind cell is new and developing (Figure 5-1). The central mature thunderstorm may become a severe thunderstorm.

Regardless of the growth pattern, supercell, multicell, or squall line, there is evidence that the internal structure of many of the severe thunderstorms is similar in each case. The severe thunderstorm develops an organized internal structure to allow it to survive longer and grow larger than the ordinary thunderstorm.

NATURE OF SEVERE THUNDERSTORMS

Simple thunderstorm models developed in the 1800s by various scientists suggested the presence of updrafts and downdrafts in severe thunderstorms as a step in explaining such observed features as lightning, hail, and tornadoes. It was only in the 1970s, however, that remote sensing techniques, specifically dual Doppler radar, was sufficiently developed to give information on the wind currents within a severe thunderstorm. These have shown such features as counterrotating vortices inside the severe thunderstorm. The existence of such a double vortex structure was suggested by the author from the theory of air flow around barriers a few years before such measurements were available for verification as will be described in the next section.

Simple visual observations of a severe thunderstorm in action provide some clues to the internal structure. Large thunderstorms extend from their cloud base at perhaps 3000 ft. to heights as great as 60,000 ft. The wind environment surrounding them typically includes light winds near their base and strong winds of perhaps 100 mph at the jetstream core near 40,000 ft. In spite of this strong shear, the large thunderstorm is able to withstand these high winds to avoid being tilted over or sheared off as happens when small thunderstorms start to grow. In addition to visual observations, satellite and other photographs of large thunderstorms show that they maintain a fairly vertical position (Figure 5-2).

The most severe thunderstorms stay in a vertical position for several hours. This tells us that the thunderstorm must have the appropriate internal wind currents to counterbalance the strong

Figure 5-2. Severe thunderstorms are able to grow very quickly through a strong wind shear environment and maintain a vertical structure that extends through the jetstream. Their tops spread out to form anvils as they penetrate into the stratosphere.

external winds which would otherwise penetrate the thunderstorm and tear it apart since it does not move forward with the same speed as the jetstream. Thus, a large persistent thunderstorm must represent a barrier to the environmental wind flow.

DEVELOPMENT OF THE DOUBLE VORTEX MODEL

Complex potential theory includes the study of mathematical functions that satisfy Laplace's equation. The solution of this equation for flow around a cylinder is shown in Figure 5-3. The question is how this solution to flow around a barrier applies to air currents inside the thunderstorm. It is obvious that the thunderstorm would form a barrier to the environmental winds if it contained a double vortex circulation as shown in Figure 5-3. A cyclonic rotating cell in the southern half of the thunderstorm and anticyclonic rotation in the northern half would produce a strong east to west flow through the center of the thunderstorm to counterbalance the environmental winds striking the thunderstorm from the west. In addition, the two counterrotating cells would cause the environmental air to flow around the sides of the storm with very little interference, since the winds on the sides of the storm are moving in the same direction as the surrounding flow.

The solution in Figure 5-3 suggested to the author and his graduate

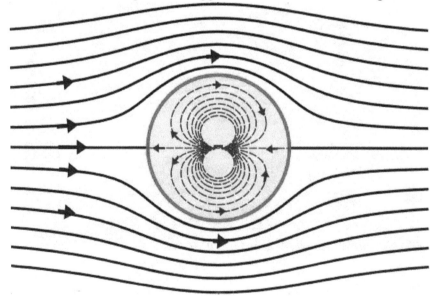

Figure 5-3. Potential theory solution to Laplace's equation for flow around a cylinder.

student, Robert Brown, that it might apply to thunderstorms. The double vortex thunderstorm model was first presented at the 1971 Seventh National Conference on Severe Storms by the author and a paper was published in the Preprint Volume.

Measurements of the internal structure of thunderstorms soon became possible from dual Doppler radar. The Doppler effect can be used to measure the speed of raindrops in the thunderstorm as they reflect radar signals back to a receiver. The use of two or more Doppler radar sets aimed at the same thunderstorm allows the three-dimensional airflow patterns within a thunderstorm to be reconstructed.

To investigate air flow inside a severe thunderstorm, radar data were requested from NSSL. The only stipulation of my request was that the thunderstorm produced a tornado. Data were received and analyzed for a thunderstorm on June 8, 1974 that contained a tornado near Oklahoma City. The Doppler data were collected as thousands of individual data points for each of the two radar sets located in different places. After analysis these were displayed as horizontal slices through the thunderstorm showing the airflow (Figure 5-4). A tornado was reported on the ground 12 minutes after the data set. The Doppler data for this storm shows the existence of the hook echo, and the presence of a double vortex structure extending vertically for a great distance through the thunderstorm. The double vortex structure was strong from 13,094 ft. (4 km) to the upper limits of the data, 23,000 ft. (7 km). A weak hook echo can be seen in the reflectivity at 6,547 ft. (2 km) that was generated by the cyclonic flow of air from the southern part of the thunderstorm that carried some of the rain around with it to give the hook-shaped radar echo. The double vortex structure of this storm was still very evident at the highest level of data, 23,000 ft (7km). It was surprising to find the double vortex in this storm but the double vortex internal flow structure of severe thunderstorms has now been verified by many other researchers including Kropfli, Segman and Diao. The double vortex is also known as vortex couplets or mesovortex couplets. Numerical modeling by Schlesinger has shown the evolution of a double vortex.

FORMATION OF THE DOUBLE VORTEX SUPERCELL

A double vortex internal structure is enhanced simply from air flowing around a thunderstorm, interacting with the sides of the storm. This action alone may be enough to create a cyclonic vortex in the southern part of the storm by the airflow against the southern edge of the

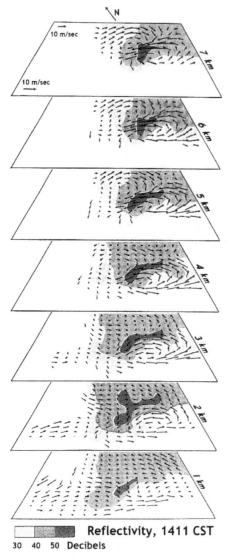

Figure 5-4. Dual Doppler Radar data from a tornado producing thunderstorm show the cyclonic circulation in the southern part of the thunderstorm at various levels. The anticyclonic circulation is also evident on the slices from 4 km to 7 km. Precipitation is concentrated between the double vortex structure at these heights as indicated by the darker shading. (From Eagleman and Lin, Journal of Applied Meteorology, October, 1977.)

storm. A more important role is played by the vertical wind shear to develop the double vortex structure as shown in Figure 5-5. As a cumulus cloud grows higher in the atmosphere it encounters stronger winds that can be observed to blow the top off small clouds. As the cloud grows higher, the extreme wind shear develops horizontal vortices, similar to roll clouds. Well below the jetstream the wind shear is sufficient to form a horizontal vortex strong enough and long enough for the ends of the vortex to tilt downward into a more vertical position. The southern end of the vortex rotates cyclonically and the northern end rotates anticyclonically as they become integrated into the thunderstorm structure. Such rotation can then be further enhanced by the airflow around the sides of the thunderstorm, thus forming a stable internal flow pattern that is compatible with the strong surrounding airflow and with the airflow suggested by solving Laplace's equation.

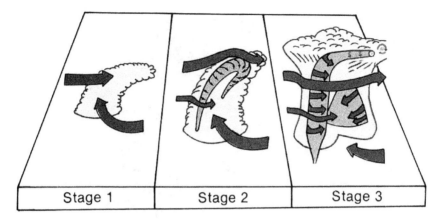

Figure 5-5. The internal structure of severe thunderstorms develops as the rising, small updraft forming the thunderstorm begins to block the surrounding air flow. The double vortex structure develops as horizontal circulation originated because of the strong vertical wind shear is tilted into the vertical. The cyclonic circulation within the southern half of a thunderstorm and anticyclonic circulation within the northern half increase further because of surrounding air-flow around the southern and northern sides of the thunderstorm.

THE MATURE DOUBLE VORTEX SUPERCELL

In the mature stage, the internal flow of a double vortex thunderstorm allows it to exist in a vertical position, since the winds inside the thunderstorm now oppose the strong environmental winds coming from the opposite direction at the upper backside of the thunderstorm with a force sufficient to counteract them. The counterrotating vortices also help draw additional warm, moist air into the front of the storm at low levels as shown in Figure 5-6. Thus, the storm has achieved internal flow patterns that allow it to survive, and, in doing so, has also created the mechanisms for the production of damaging surface winds, lightning, hail, and tornadoes. Thunderstorm models such as the double vortex model, allow us to consider the thunderstorm as a whole; something that is not always possible from other approaches.

A severe thunderstorm is not generally stationary but is moving. As the strong environmental winds flow against the back and around the sides of the thunderstorm, they propel the thunderstorm forward and create a wind at low levels that strikes the thunderstorm from the front (northeast side) as previously explained by relative winds. This air

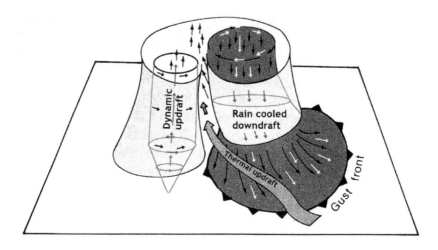

Figure 5-6. The cold air outflow through the rain area in the northeast part of a thunderstorm spreads out ahead of the storm as it strikes the ground. This generates a gust front that is somewhat similar to a miniature cold front. This helps to push the warm, moist, inflow air current into the thunderstorm.

goes into the thunderstorm and rises, since it is warm, humid air that is light and buoyant. It rises because of its temperature and water vapor content, until it strikes the back side of the thunderstorm, where some of it starts to spiral around cyclonically on the southern side or anticyclonically on the northern side of the cell. The darkest areas in Figure 5-4 correspond to the greatest water concentration and this shows how it is pulled into the storm between the vortices.

Raindrops begin condensing within the interior of the storm as the warm air rises within the central part. Most of the rain falls from the northeast section of the thunderstorm, since the upper level winds blow the raindrops downwind into the anticyclonic rotation. The anticyclonic rotation within the thunderstorm does not develop as much strength as the cyclonic vortex because of the obstruction of large masses of rain. The anticyclonic vortex is also weaker than the cyclonic vortex because of the large-scale cyclonic curvature of the jetstream. As cyclonic vorticity and curvature are fed into the cyclonic vortex, also called **mesocyclone**, its rotation becomes much more pronounced than the anticyclonic rotation. This leaves the southern part of the thunderstorm with a very strong counterclockwise or cyclonic rotation that is further enhanced as it develops strength enough to centrifuge raindrops out of the vortex to eliminate their obstruction to airflow. As the cyclonic vortex grows it begins to simulate a vertical tube extending from the lower part of the thunderstorm to its top. The interior of the vortex tube then develops an updraft because of the higher surface pressure and the influence of the jetstream flowing at the top of it. This **dynamic updraft** through the center of the cyclonic rotation, combined with the horizontal rotation, eventually extends down to the ground as a tornado. The anticyclonic vortex does not produce a tornado because it is weaker as just described.

The outer shell of the cyclonic vortex is generated by the **thermal updraft**, the warm, moist air traveling from low levels into the front part of the storm that rises and collides with the environmental winds at the back of the thunderstorm, thus feeding the cyclonic vortex. Within the cyclonic vortex, the smaller dynamic updraft becomes quite strong due to the pressure difference between its top and bottom. The core of the anticyclonic vortex does not have a central updraft because it is filled with rain and the effect of falling rain is great enough to initiate a downdraft through its core. The major rain area, therefore, occurs in the leading northeast part of a thunderstorm, while if there is a tornado, it extends from the southern part of the cloud.

It has been observed that once a thunderstorm has developed the appropriate internal structure to generate a tornado, it approaches more steady-state conditions and has a longer lifespan. Also, because of the more steady-state internal structure, a thunderstorm that develops one tornado is more likely to develop another one after the first one dissipates.

The downdraft within the rain area of the anticyclonic vortex is typically strong enough to reach the ground. Sometimes it is strong enough to create a microburst. The **wet microburst** is 2 miles or less in diameter and may cause extensive damage at the surface. It is the first blast of rain cooled air and lasts for only a short time. Ordinarily the downdraft from inside the anticyclonic vortex is only a cool wind that strikes the ground with less velocity and spreads outward ahead of the approaching rain in the leading part of the thunderstorm to form a gust front. This gust of cooler air helps feed the thermal updraft into the storm by starting the warm surface air upward as seen in Figure 5-6. This makes the thunderstorm even more efficient after it reaches the supercell stage. Thus, the air currents reinforce each other, with additional reinforcement as raindrops fall through the

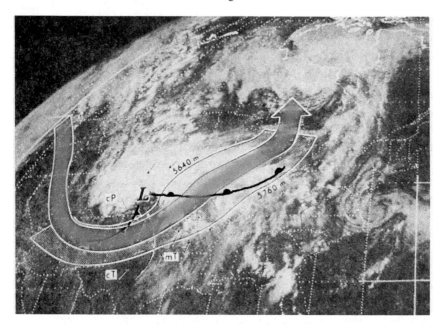

Figure 5-7. Thunderstorm movement is frequently from the southwest because of the upper air currents, as shown here for June 8, 1974, at the time of tornado formation in Kansas and Oklahoma. The cloud in central Oklahoma was analyzed in detail by dual Doppler radar as shown previously. The large arrow shows this typical jetstream flow pattern from the southwest. (From Eagleman and Lin, Journal of Applied Meteorology, October, 1977.)

cloud and evaporate some water to cool the air and contribute further to the downdraft, since cold air is denser and accelerates downward.

THUNDERSTORM MOVEMENT

The movement of thunderstorms is determined primarily by the upper level winds with some influence from thunderstorm rotation. If the thunderstorm is not rotating, and the average winds from the surface to the top of the thunderstorm are from the southwest, then the thunderstorm will move with the prevailing winds. A typical tornado weather pattern is shown in Figure 5-7. It corresponds to the dual Doppler radar data, shown previously. This figure shows a typical upper airflow pattern with the jetstream coming from the southwest and a low-pressure center located just east of the trough and slightly north of the jetstream.

The rotation of a thunderstorm will affect its direction of travel if the rotation is very pronounced. Rotation of a thunderstorm would be expected to affect its motion in exactly the same way that a baseball pitcher puts spin on the ball to make it curve. Cyclonic rotation produces thunderstorm movement to the right, when looking downstream with the mean wind, whereas anticyclonic rotation produces movement to the left. If the mean wind is from the west, and a large thunderstorm has considerable cyclonic rotation, then the moving force is the average wind from the surface to the top of the thunderstorm, which is carried along with the winds, except for some deviation to the right of the mean wind (Figure 5-8). This rotation force is called the **magnus force**, and it deflects a thunderstorm just as a baseball pitcher spins the ball to throw a curve ball. The flow of air around the spinning thunderstorm produces greater pressure on one side, and this causes it to curve.

Since most severe thunderstorms have some cyclonic and some anticyclonic rotation, the rotations tend to cancel except that the cyclonic rotation is usually much stronger, resulting in thunderstorm movement to the right of the mean wind. Most severe thunderstorms come from the southwest in the direction of the average wind, with some deviation to the right depending on the strength of the cyclonic rotation.

Thunderstorms with considerable internal rotation are pushed along

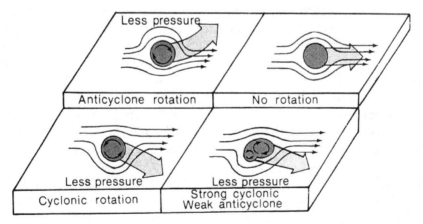

Figure 5-8. Thunderstorm movement is influenced by the upper-level winds that carry it along and by its rotation. Anticyclonic rotation is accompanied by deviations to the left, while cyclonic rotation causes it to travel to the right of the average air stream. The reason for these deviations is a reduction in pressure on one side of the thunderstorm because of the rotation.

at a slower rate than those with less rotation, since a well-developed double vortex structure forces the approaching winds around the outside of the storm with minimum friction and a corresponding loss of power to push the storm along. If the thunderstorm rotation is at a greater speed than the relative wind around it, the storm can move upstream for a short time period as the greater rotational winds propel it against the surrounding winds. This can continue for only a short time, since the thunderstorm must derive its energy for rotation from the surrounding air stream, but is possible particularly when the winds are changing and the rotation within the thunderstorm has reached its peak. In general, however, the thunderstorm moves along at a rate determined by the environmental winds with a typical speed of 25 mph. It has been observed that most severe thunderstorms slow down at about the time they form a tornado. This behavior is intriguing, as if they were stopping to liberate the tornado. Actually, this is what is happening since the increasing circulation, which generates the tornado, also reduces the ability of the environmental winds to push the thunderstorm along.

SPLITTING THUNDERSTORMS

Another interesting observation of severe thunderstorms is that occasionally a single thunderstorm will split into two different cells. When this occurs, the southernmost cell turns toward the right, while the northern one turns toward the left as shown for several different splitting storms in Figure 5-9. This is another verification of the prevalence of the double vortex internal structure of thunderstorms with cyclonic rotation in the southern part of the cell and anticyclonic rotation in the northern part. Such an internal structure is the only explanation for the observed paths of the storms after splitting because the paths they take reveal their circulation.

The splitting may happen repeatedly with one cell splitting into two cells, then several hours later, these cells split again. Storm B in the figure was the anticyclonic half of the first split and it must have added the cyclonic rotation to become a double vortex thunderstorm within 2.5 hours because it split a cyclonic half that deviated to the right that headed toward Oklahoma City.

This storm is also interesting in that it was the cyclonic half of a split thunderstorm and either redeveloped the anticyclonic half to form another double vortex thunderstorm within 2.5 hours or a new double

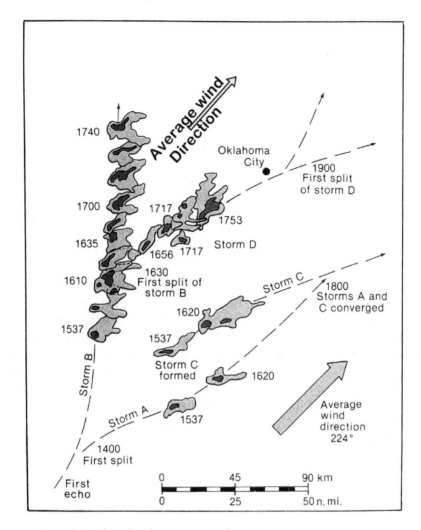

Figure 5-9. When thunderstorms split the cell on the right deviates to the right of the average winds while the one on the left deviates to the left of the average wind. The only plausible explanation for this is the double vortex internal structure of the thunderstorm. (From NSSL Tech. Memo 41)

vortex thunderstorm formed from its ruminants because the storm split again after it passed Oklahoma City into two storms deviating to the right and left of the average wind. It is also interesting that a single thunderstorm can generate several additional thunderstorms that are usually severe and can produce tornadoes.

OTHER THUNDERSTORM FORMATION FACTORS

Although thunderstorms form from the appropriate atmospheric conditions there is some weak evidence that points to the existence of preferred paths for thunderstorms. One of these seems to be toward large cities such as Kansas City or Oklahoma City. The localized area where thunderstorms originate may be unique with a topographic feature such as a hill or darker fields that could help initiate upward movement of air sufficiently to trigger the development of a thunderstorm. This then generates thunderstorm development in the same spot time after time. There are indications of this if you look at tornado paths across Arkansas, for example, where there are defined ridges with higher topography. The paths of tornadoes are parallel to many of these features although this does not prove a cause and effect relationship.

There may be some effect from cities themselves on thunderstorm formation and tornadoes. We have measured the urban heat island in and around cities of various sizes and found that it is related to city size. For example, the heat island average from many measurements for Lawrence, Kansas when its population was 50,000 was 3.6°F. A heat island of 5.4°F exists in Topeka (population 125,000), and a heat island of 7.2°F was the average in Kansas City (population more than one million). If the air over the city is hotter, updrafts would be favored over the city because of the lighter more buoyant air. Thunderstorms approaching a particular city with rising air above it, would have some tendency to move toward the rising air and warmer temperature. The heat island would not be expected to play a major role in thunderstorm formation and movement but could be a minor factor.

NATIONAL SEVERE STORMS LABORATORY (NSSL)

The National Severe Storms Laboratory is a federal research laboratory where severe weather is investigated including weather radar, tornadoes, flash floods, lightning, damaging winds, hail, and winter weather. NSSL is located in the National Weather Center (NWC) in Norman, Oklahoma. NSSL serves to enhance NOAA's capabilities to provide accurate and timely forecasts and warnings of hazardous weather events. They accomplish this mission through research to advance the understanding of weather processes, research to improve forecasting and warning techniques, and development of operational applications. NSSL transfers

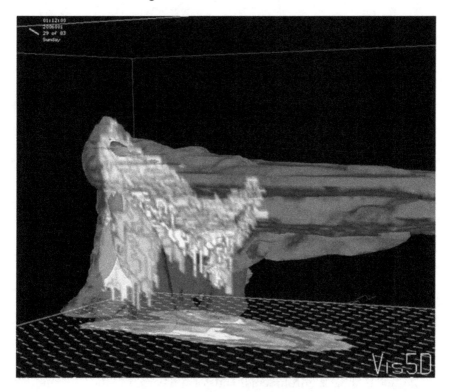

Figure 5-10. The NSSL computer supercell thunderstorm model. (Photograph from NSSL website)

new scientific understanding, techniques, and applications to the Storm Prediction Center (SPC) of the National Weather Service (NWS).

Some of the research there aims to better understand when and where severe weather will occur, by studying thunderstorms through direct observation in the field or by making computer simulations. They apply this knowledge as they develop and enhance weather prediction models and techniques to support the NWS mission to provide weather forecasts for the U.S to SPC so that the public has the benefit of current severe weather forecasting techniques. One of the methods used to study severe thunderstorms is computer modeling. An image of the supercell computer model is shown in Figure 5-10. Researchers there have created the computer model that simulates thunderstorms in 3-D. This model is used to study how a thunderstorm behaves when encountering different weather conditions. This work has led to improved forecasting and warnings, and has increased our understanding of environments that are favorable for the formation of thunderstorms.

A very difficult problem in forecasting is predicting exactly when and where a thunderstorm will form. Researchers there are engaged in a Convective Initiation Project to use computer simulations to study what precursors signal a developing thunderstorm. Collaborators on the Warn-on-Forecast project are working to create forecasts of severe weather so specific that forecasters will be able to issue a warning based on that forecast before the weather forms – up to an hour or more in advance. To do this, they are working on techniques to move from the practice of "warn on detection" to "warn on forecast" to extend warning lead times.

SUMMARY

Severe thunderstorms produce frequent lightning, and locally damaging winds or hail. They may also contain one or more tornadoes. A severe thunderstorm develops as a single supercell thunderstorm, squall line thunderstorm, or multicell thunderstorm. In each case, the severe thunderstorm generates an organized internal structure that allows it to last longer and grow larger than the common variety of thunderstorms.

Large thunderstorms are observed to grow vertically through strong wind shear environments, and form a barrier to the surrounding airflow. Air currents form a double vortex inside the severe thunderstorm with a cyclonic vortex in the southern part, and an anticyclonic vortex in the northern part of the thunderstorm. As the cyclonic vortex dominates, due to general cyclonic curvature of the atmosphere and obstruction from rain in the anticyclonic vortex, it begins to simulate a tube extending through most of the thunderstorm. Its extension to the surface produces a tornado.

The movement of severe thunderstorms is in response to the mean winds through the atmosphere with influence from the rotation of the thunderstorm. A strong double vortex structure allows the thunderstorm to slow down because of less friction as the airstream flows around it. If the cyclonic rotation is stronger than the anticyclonic, the thunderstorm will move to the right of the mean winds. Thunderstorms that split into two cells travel in separate directions, with the southern cell moving to the right and the northern cell to the left of the mean wind. This movement can only be explained by the double vortex structure of a severe thunderstorm.

Severe thunderstorms typically last for an hour or two with the longest lasting storms only a few hours. They may follow the same path more than once because of unique topographic features.

NSSL researchers use computer models to study the formation of severe thunderstorms with the objective of improving severe weather information that is supplied to the Storm Prediction Center. This center then produces the forecasts of severe weather that is eventually relayed to the public.

You can see a video of the internal flow characteristics including the double vortex on YouTube at:
https://youtu.be/uV-MboO_fr4

Joe R. Eagleman

6

The Strongest Storm on Earth

As we drove through the blinding rain, we both sensed that we should have gone to our basement instead of chasing the storm. As we broke out of the rain area traveling southward, we saw the tornado ahead of us and to our right. We thought that perhaps we could cross in front of it and circle behind out of danger. Our plan seemed to be working, and gave us a view that was overwhelming as we watched houses reduced instantly to bits of wood, lights flashing as electric lines snapped, and cars sailing through the air. Our feelings of triumph in maneuvering around the tornado were short-lived, as the roar of a second funnel drowned out the sound of our car. Before we could react, we were sailing through the air up into the second tornado. As the tumbling car gave us an upward view we looked into a long smooth-walled tube lighted by glowing electricity. As we began to descend, we saw buildings and trees zipping past at speeds that we knew would mean instant disaster if we struck them, but we were powerless to change our course. Just as we thought we were finished; we were again lifted upward. Before we knew what had happened, we were deposited, as gently as we had been lifted, on top of a flat roof. As we found stairs that led us to the kitchen of a restaurant, we encountered some very surprised cooks who saw us come from the door to the roof. We said, "We'll be back for the car tomorrow", and left them almost as bewildered as we were.

Yes, true stories like this one from an Oklahoma tornado are often stranger than fiction.

TORNADO DEVELOPMENT

Tornadoes are produced as a thunderstorm develops an organized internal structure of sufficient strength to extend the vortex from the cloud base to the ground. The severe thunderstorm that develops a tornado is normally the largest thunderstorm in a squall line or a very large isolated supercell. Supercells grow to great heights and extend from near the surface through the jetstream, where the interaction there helps provide energy for maintaining an organized internal structure that can support a tornado as in Figure 6-1. This organized structure has been considered previously, but will be reviewed as related to tornado generation.

The warm, moist air of the thermal updraft flows in between the double vortex structure at low levels and rises up through the storm to hit the back side and oppose the propelling winds that are pushing

Figure 6-1. Large tornadoes are generated by appropriate air currents within a thunderstorm. This photograph of an oil painting by the author shows the east side of a severe thunderstorm with a tornado to the south and rain area to the north.

the storm along toward the northeast. The thunderstorm interacts with the jetstream, which provides a suction over the top of the storm, especially over the cyclonic vortex that has become better developed because of large scale cyclonic curvature within the jetstream. The cyclonic vortex is stretched downstream by the jetstream. It then forms a link between the cloud base and upper atmosphere by providing a tubelike connection up to the jetstream level. The dynamic updraft then develops through the core of the cyclonic vortex, or is reinforced, to support the rotation of the cyclonic vortex. This generates a vortex tube of sufficient strength to reach the ground. This, then, is a tornado.

Many factors, therefore, are involved in arriving at a steady-state structure within a severe thunderstorm capable of tornado production. Major factors are the thermal updraft, which contains the unstable, warm, moist air; the jetstream; and the dynamic updraft through the mesocyclone or rotating southern part of the thunderstorm.

The generation of the double vortex structure is initiated by wind shear and then strengthened by the jetstream and intensified by airflow around the south side of the thunderstorm, which adds circulation to the cyclonic vortex, while similar flow around the north side of the thunderstorm contributes to increased anticyclonic circulation. In order to develop a tornado, the cyclonic vortex must extend down through the cloud base.

The thunderstorm acts as a unit with the vortices generated in the upper part of the thunderstorm extending downward toward the cloud base where the winds at this lower altitude are moving in an opposite direction to the winds higher up. This has the effect of concentrating incoming air at the cloud base much as if a suction were applied in front of the thunderstorm that pulls the warm, moist air in between the vortices to form the thermal updraft. The whole thunderstorm structure is then maintained and becomes more efficient.

As condensation occurs, additional heat is provided to the thermal updraft and more buoyancy is created adding further to the wind speeds. The cyclonic vortex becomes even more efficient as the circulation becomes strong enough to centrifuge the raindrops outward, thus eliminating their obstruction effect. Lightning discharges within the mesocyclone and the thermal updraft add further to the buoyancy of updrafts (Figure 6-2). As the large-scale atmospheric environment and the internal circulation within the thunderstorm operate in harmony, the mesocyclone becomes strong enough to support the smaller, but much more intense, circulation of its attached

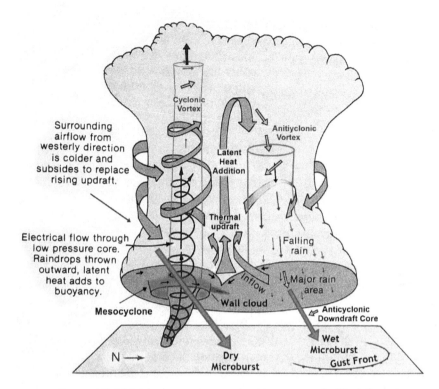

Figure 6-2. Many factors act in unison to form a tornado. The inflow of warm moist air in the front of the thunderstorm at low levels opposes the higher-level southwesterly flow to initiate a large cyclonic vortex. Electrical flow and latent heat contribute to its strength after its formation.

tornado. As we have previously seen, the high wind speeds within the tornado are possible because of the conservation of angular momentum.

Much smaller tornadoes are formed by thunderstorms that do not have the advantage of all the appropriate atmospheric conditions. Thus, the smaller thunderstorm over the Gulf of Mexico frequently generates waterspouts from the strength of the thermal updraft, combined with much smaller side winds, without the benefit of the jetstream. Similarly, smaller tornadoes develop and last for a short time in areas where severe weather has not been forecast, as indicated by older statistics that showed only about 30% of tornadoes forming in forecast tornado watch areas, while about 50% of the larger, more destructive tornadoes formed in forecast areas. It is encouraging to note that these statistics are constantly improving so that today most of the tornadoes occur in forecasted areas.

TORNADO DAMAGE PATHS

Tornadoes originate within the mesocyclone when the horizontal circulation and updraft core become strong enough to extend below the cloud base. The type of tornado damage path depends on where the tornado is generated within the mesocyclone. If it is developed in the center of the mesocyclone, and if the thunderstorm is moving at a fairly high rate of speed, then the tornado damage path will be long and straight (Figure 6-3). If the tornado develops at the southern edge of the mesocyclone, which occurs often, the mesocyclone carries the tornado around with it in a cyclonic direction. When the tornado passes the front central part of the thunderstorm and encounters the rain area it soon dissipates while another funnel forms on the southern edge of the mesocyclone. This results in a series of tornadoes. It is not uncommon for three or four tornadoes to be produced by thunderstorms of this type. If the thunderstorm generates one tornado in the center of the mesocyclone, it normally lasts longer, although it may originate others on the edge of the mesocyclone as well. If the tornado forms in the center and the thunderstorm moves at a slow rate of speed, the tornado will produce a looping path on the ground and may also last for a considerable length of time.

The three major types of damage paths of tornadoes are, therefore, the repeating type, in which a single thunderstorm produces 2 to 4 different tornadoes, one after the other; the looping type; and the straight type. The looping damage path, like that of the Lubbock tornado in 1970, occurs from a tornado in the central part of the mesocyclone with a thunderstorm that is moving quite slowly due to light upper level winds. The tornado takes a looping path because of the uneven nature of the earth's surface and the difficulty of taking air into the tornado core from all sides equally. If the tornado originates in the central part of the mesocyclone and the thunderstorm is moving at a fairly rapid speed, it produces a straight path. The repeating damage paths are produced as tornadoes develop on the southern edge of the thunderstorm within the cyclonic circulation of the thunderstorm, which carries the tornado around to the front part of the storm and then into the middle where it dissipates. As it dissipates, the same conditions that caused the first

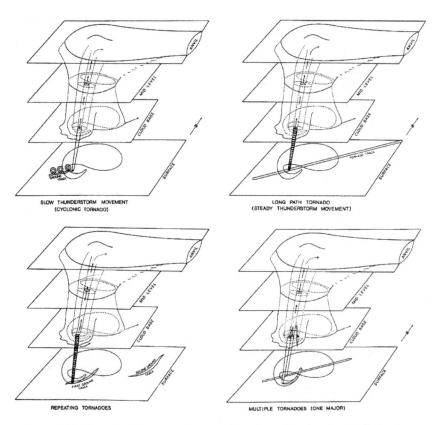

Figure 6-3. Tornado damage paths are related to the speed of travel of thunderstorms and to their location within the mesocyclone. Long path tornadoes are more centered within the mesocyclone in the southern part of thunderstorms, while intermittent tornadoes are off-centered with respect to the mesocyclone. (From Eagleman, Muirhead, and Willems, Thunderstorms, Tornadoes, and Building Damage, D.C. Heath & Co., Lexington, Mass. 1975.)

one creates another.

An analysis of Missouri tornadoes by Grant Darkow showed that about one-third of the tornadoes were produced by parent clouds that developed more than one tornado. These tornadoes did about half the total damage because the path lengths were a little longer than normal. An analysis performed to see if a constant interval occurred between the individual tornadoes gave some indication of a repeated time interval. Most thunderstorms that formed more than one tornado developed the second one between 30 and 60 minutes after the formation of the first. Very few developed a second tornado after 90 minutes had elapsed since the first tornado. The most common interval between tornadoes was 45 minutes with three or four typically generated from one parent cloud.

Tornadoes may stay on the ground for several hundred miles. In 1927, a tornado in Kansas was on the ground for 105 mi. The Tri-State Tornado, in 1925, developed in Missouri, and went through Illinois and Indiana, staying on the ground for over 250 mi.

The typical tornado damage path is only about 5 mi. in length and 150 yards in width. The typical speed of a tornado is 30 mph. The fastest ones travel at speeds of 100 mph or more as they change form. The tornadoes in the major outbreak on April 3, 1974, traveled at forward speeds of 60 mph because of the very strong jetstream. If the tornado damage path is typical, 5 miles long, and the tornado travels at a speed of 30 mph, it is on the ground for only 10 minutes. This illustrates the difficulty of getting warnings to the public because of the short lifetime of the event.

PRESSURE, WIND SPEED AND EF SCALE

Many of the characteristics of tornadoes are still largely unknown. For example, the minimum pressure within a tornado is not well known in spite of its importance. Tornadoes have passed over weather stations where barometers were in operation. However, the response time of barometers, as well as the time required for the equalization of pressure inside and outside the buildings, prevents the measurements from being very meaningful. The pressure inside a tornado is low enough to cause buildings to explode as the pressure suddenly decreases when the tornado arrives. We can only guess at the minimum pressure within a tornado. It could be, theoretically, as low as the pressure at the top of the vortex tube that extends from the surface to the 300 mb level. This means it could be as low as 300 mb if the vortex tube were spinning fast enough to prevent any air from getting in from the outside.

Similarly, we do not have measurements of the wind speed within tornadoes. As the tornado passed over the weather station at Topeka a piece of debris hit the anemometer before it reached 100 mph. On another occasion in Springfield, Missouri, a tornado that passed over a weather station spun the three-cup anemometer so fast that one of the cups blew off before it reached 100 mph. The best estimates are obtained by looking at tornado damaged structures and estimating the amount of force required to bend particular strips of metal, for example.

Figure 6-4. Extreme wind speeds in tornadoes generate some unusual results. Wood is a softer material than glass but the winds of the Lubbock tornado (EF5) carried the splinter shown above at such speed that it was able to penetrate this windshield even though the splinter struck the windshield at a slanted angle.

The light poles at Texas Tech University in Lubbock, were designed to withstand wind speeds of 155 mph, but the tornado in 1970 toppled them over. Wind speeds great enough to force straws into poles, pluck feathers off chickens and do other strange things, commonly occur in tornadoes (Figure 6-4). An analysis of structural damage in the Dallas tornado of 1957 indicated that winds in excess of 300 mph probably occurred. Wind speeds in tornadoes occur over a wide range. Because of the lack of measurements, wind speeds were estimated by Fujita from surveying the type of damage caused by a tornado. The original Fujita scale has been modified by the National Weather Service to be called the Enhanced Fujita Scale (EF).

This scale that became operational on February 1, 2007 is used to assign each tornado a rating based on the estimated wind speeds determined by damage. When tornado damage is surveyed the amount of damage to various types of buildings and objects is used to assign a rating from EF0 to EF5. Estimates of the ability of different types of structures to withstand winds of a particular speed allow wind speed estimates to be assigned to each category.

EF-Scale	Intensity	Wind Sp. mph	Damage
EF0	Light	65-85	Branches broke
EF1	Moderate	86-110	Roof damage
EF2	Significant	111-135	Roofs off
EF3	Severe	136-165	Walls torn off
EF4	Devastating	166-200	Houses leveled
EF5	Incredible	Over 200	Trees debarked

APPEARANCE

A tornado is visible because of dust and debris picked up from the ground and from the condensed water vapor. Water vapor inside the funnel condenses because the pressure is less. The lower pressure inside the funnel causes the air to expand and cool with the result that water droplets condense. Figure 6-5 shows some of the different types of

Figure 6-5 Some of the different shapes and sizes of tornadoes. (Composite of NOAA photographs).

funnels. Those on the left have debris or dust in the lower part of the funnel and condensed water vapor in the upper part to make them visible. The funnel may vary from massive as on the right middle photograph to small rope shape as in the top right photograph.

A few eye witness reports of close encounters with tornadoes are available. Milton Tabor was a student in Lincoln, Nebraska, on March 23, 1913 when a tornado originated right over his head. He describes it in *Weatherwise,* April, 1949: "The tornado cloud formed while we were enjoying a picnic and groaned furiously high in the air straight over our heads. We looked up into what appeared to be an enormous hollow cylinder bright inside with lightning flashes and black as night all around. The noise was like ten million bees, with a roar that defies description".

The condensed water vapor can produce a white tornado or dark depending on the background. With the rain area behind it as in the top right photograph in Figure 6-5 it appears white. When the clear sky is behind it, the tornado appears darker. A rare photograph of two tornadoes from the same cloud is shown in Figure 6-6. These are both white because of the dark rain in the background.

Figure 6-6 Two white tornadoes from the same storm formed over the Great Plains. (NOAA Photograph).

DISTRIBUTION AND NUMBER

Tornadoes are most common in the month of May and June (Figure 6-7). In general, the winter months have few tornadoes while the spring months from April through June have many more. The number of tornadoes that occurs each year averages 120 in Texas alone, while Oklahoma has 58, and Kansas is third with 48 tornadoes. If the number of tornadoes per unit area is used, Oklahoma becomes the greatest tornado producer. These results are based on statistics gathered since 1956; prior to this time the records are not as accurate

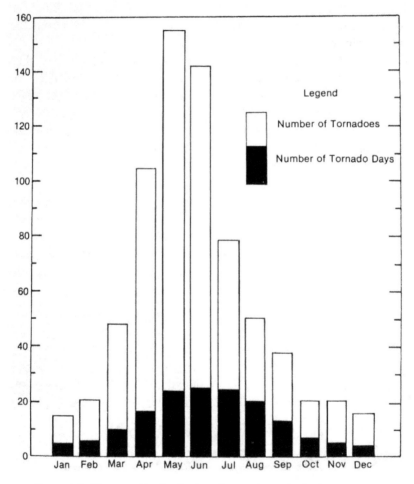

Figure 6-7. The monthly distribution of tornadoes shows that more occur in May and June than other months. This distribution is based on 18,641 tornadoes. (NOAA data.)

since the number of tornadoes per year increased because of improved reporting. Kansas, in spite of Wizard of Oz fame, is third in both total number of tornadoes per year and number per unit area. Surprisingly, Florida is fourth in both categories followed by Iowa and Missouri, on a unit area basis.

More tornadoes occur in the Central United States than anywhere else in the world because of the unique combination of weather conditions and topography. The mountains to the west of the Great Plains cause the air flowing over them to subside as it flows into the Great Plains. This creates upper air inversions that affect cumulus cloud development as was shown in Figure 4-5. Surface weather patterns frequently include a high-pressure area off the East Coast of the United States in the Atlantic Ocean. This brings south winds from the Gulf into the Great Plains area (Figure 6-8). The lower atmosphere frequently has very hot, humid air from the south winds at the surface with cooler, drier airflow above from the west as the air flows over the mountains. This unique combination sets up the proper atmospheric conditions for severe thunderstorms and tornado generation as previously described.

The number of deaths from tornadoes in different states does not match the total number of tornadoes. Tornadoes cause more deaths

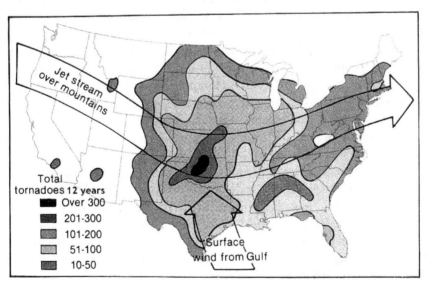

Figure 6-8. The central United States is the favored location of tornadoes as the Rocky Mountains block surface winds. The jetstream flows over them while warm surface air is fed into the Central Great Plains.

in southern states, such as Mississippi and Alabama, than in other locations. Midwestern states that have more tornadoes have fewer deaths; Kansas and Oklahoma, for example, average only seven deaths annually in spite of the greater number of tornadoes there. This indicates that preparation and warnings are important in protecting people from the storms.

The year 1973 was memorable for tornadoes. A total of 1109 tornadoes occurred in 1973, to break the previous record for the most tornadoes in any year, set in 1967 with 929 tornadoes at a time in history when the average number of tornadoes per year was 700 per year. A new record was set in 2011 as it was the busiest for tornadoes in the U.S. since recording began in 1950, when 1,894 tornadoes were reported, according to NOAA's Storm Prediction Center (SPC). Much better reporting of tornadoes since the 1970's has increased the average number of tornadoes per year to 1200.

Tornado extremes of a different type were established in 1974. A super outbreak of tornadoes occurred with a record number of tornadoes, fatalities, and injuries. Although the total number of tornadoes for the year was quite high (945), 144 of these occurred within a two-day period, April 3 and 4. A total of 361 people were killed by tornadoes in 1974 and 6,915 more were injured. People were killed in 20 different states by 64 different tornadoes. Figure 4-7 showed some of the hook echoes. Alabama had more deaths than any other state with 79 fatalities, but the single largest killer tornado occurred at Xenia, Ohio where 34 people were killed on April 3. Most of the deaths in Alabama (77) were during the April 3 and 4 outbreaks. Thirteen different states had major tornadoes during this outbreak. Eleven of these states suffered fatalities as a result of the tornadoes during this period. This super outbreak of tornadoes on April 3 and 4 caused 315 deaths and 5,514 injuries with over half a billion dollars damage to property.

This super outbreak in 1974 still holds the record for largest number of strong tornadoes with 30 EF4 or greater tornadoes. The number of deaths from this outbreak has only been exceeded once by the tornadoes on April 27-28, 2011 with 324 deaths.

The time of year with the greatest threat of tornadoes varies across the United States. The tornado threat starts earliest in the southeast in April for many of the states. The time of maximum threat moves northward with the seasons and is greatest in May and early June for a band extending eastward from the central United States. Tornado activity in July is more likely to be in the north central and northeast.

TORNADO DETECTION

It is the responsibility of NOAA's Storm Prediction Center (SPC) to issue tornado watches based on such atmospheric factors as the jetstream, upper air divergence, presence of a squall line, and other factors as previously described. Such tornado watches are issued for large geographic areas covering several hundred square miles. Tornadoes, of course, only affect a very small portion of this area.

A tornado warning indicates that a direct threat of tornadoes exists in a localized area. Specific warnings are provided by the local weather service, Civil Defense Director, and television or radio station operators. They are issued on the basis of the observation of a hook echo (Figure 6-9) or other signatures on the radar screen or from eyewitness reports, normally verified by highway patrolmen. If the local weather station is equipped with radar, the formation of a hook shaped echo is the usual basis for issuing a tornado warning. The hook shape on the screen only indicates that the thunderstorm has generated a mesocyclone which may be capable of producing a tornado. More than half of the tornadoes

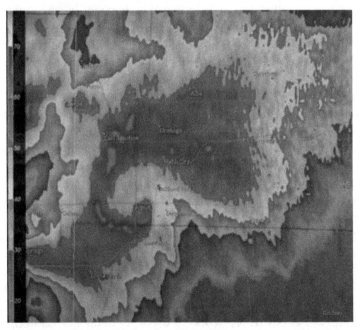

Figure 6-9. The hook echo for the devastating Joplin tornado May 22, 2011. It killed 158 people and caused damages of $2.8 billion. (NOAA Photograph)

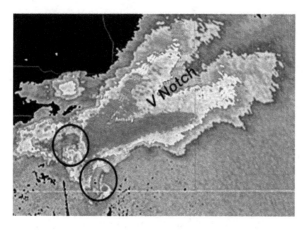

Figure 6-10 The v-notch for the Oklahoma supercell April 14, 2012 along with the expected location of the double vortex. (NOAA Photograph)

occur from thunderstorms that do not produce a hook shaped echo on the radar screen. In order to see the hook the radar must be aimed at the lower one-fourth of the thunderstorm. Thus, the range of the radar is limited for severe thunderstorm detection because of the earth's curvature.

The hook echo forms as the radar displays the reflection from raindrops within a horizontal slice through the lower part of a severe thunderstorm. The radar signal is reflected back to the station by raindrops carried in the outer part of the mesocyclone. This circulation brings rain originating in the thermal updraft around to the front part of the thunderstorm forming the hook shape on the radar screen. Most thunderstorms that form a hook echo also develop a tornado, although the tornado does not show up on a radar screen.

Another feature that is useful for identifying supercells that are likely to contain a tornado is the v-notch or sometimes called the flying eagle. An example is shown in Figure 6-10. The notch that sometimes shows up on supercells can be explained by the thunderstorm blocking the upper air flow, especially if it has a double vortex structure. The expected location of a double vortex inside the thunderstorm that reaches to the jetstream is shown in this figure. Without a double vortex to block the environmental air stream it is not likely that the blockage would be sufficient for a notch to be developed. Some of the rotation within both the cyclonic and anticyclonic vortices would be tilted into the horizontal by the force of the jetstream. This would cause a separation downstream creating a notch.

Researchers at NSSL are working on other radar detection options. One of them is the development of a Mesoscale Detection Algorithm to identify the mesocyclone of a thunderstorm. Another effort aims at identifying a Tornado Vortex Signature. This is a Doppler radar velocity pattern that can be seen a few miles above the ground before the tornado touches the ground. It is a smaller circulation than the mesocyclone and may be a good indication of a tornadic circulation reaching the ground. These efforts would be very helpful in getting tornado forecasts to the public quicker.

Many communities have sirens located throughout that are activated during a tornado warning Figure 6-11. The tornado sirens that are operated as a warning device may not cover the entire community very effectively. The sound from sirens is not likely to penetrate to people inside houses during severe weather with lightning, thunder, and high winds. Thus, the use of the specially designed radios is a trend of the future. The National Weather Service broadcasts weather information continuously at frequencies of 162.4 to 162.55 MHz from a large number of stations over the United States.

Figure 6-11 Tornado siren.

Cell phones are very useful in keeping a close watch on the weather. Apps are available from The Weather Channel and others that provide not only the forecast but radar close to real time. Once a severe thunderstorm has developed its movement can be tracked on the cell phone and its path toward or away from your location can be monitored. In addition, the hook echo may be seen if it is present.

Individual warnings may be provided by the noise of a tornado. The noise has been variously described as ten million bees, many freight trains or jet planes. The noise is so loud that it may provide a warning in itself but response time would be very short. The source of the sound is uncertain, but it could be the result of supersonic winds within the thunderstorm.

The appearance of the sky may indicate tornado formation. A "tornado sky" is shown in Figure 6-12. Such clouds form when the atmosphere is very unstable and are usually present underneath the

Figure 6-12. A tornado sky (mammatus clouds) such as shown here indicates severe turbulence in thunderstorms and usually is present when tornadoes occur. Some thunderstorms, however, have this mammatus formation and never form a tornado.

anvil of a supercell that has a tornado but their formation does not always mean that a tornado has formed.

Animals may also provide tornado warnings to people. Dogs, for example, hear sounds at ranges beyond human capabilities (the principle behind silent dog whistles). There are indications that animals can also hear tornadoes at greater distances than people. Dogs may drastically alter their behavior when a tornado is approaching. In one account, a dog that never came inside the house scratched on the door and come in when the door was opened and crawled under a bed as a tornado approached.

TORNADO SAFETY

The chances for surviving a major tornado can be improved greatly by advance preparation and proper response during a tornado warning. Advance preparation will be considered in a later chapter dealing with building design and construction techniques. A person's immediate response to a tornado threat should be governed by a good understanding of tornado safety factors.

The safest location in a tornado-stricken area is within underground concrete shelters. Rooms or storm shelters constructed of at least 10 in. of reinforced concrete for the walls and ceiling are ideal shelters. Basements offer considerable protection from tornadoes if they are constructed of reinforced concrete. Damage surveys have shown that safest locations are opposite the approach direction of the tornado; thus, the northeast rooms are safer.

Safest locations in houses without basements are in the northeast rooms (rooms opposite the approaching tornado) with smaller rooms and closets safer than large rooms. Boards and debris frequently penetrate the south and west walls and these walls are also more likely to fall inward than others.

Automobiles offer little protection from a tornado if they are caught in it. Many of the deaths during the Wichita Falls tornado occurred in automobiles. The people caught in an Oklahoma tornado described in the opening paragraph of this chapter were extremely lucky. I never heard how they got the car off the roof. A large tornado traveling at 50 mph is not as easy to maneuver around as might be expected, especially if traffic is heavy. Mobile homes are not a safe location during tornadoes. Mobile home parks should provide a community shelter, perhaps an underground recreational area, for use by residents.

If you are caught in open country during an approaching tornado seek a small depression such as a ditch running perpendicular to the path of the tornado. This may be helpful since winds and debris traveling at speeds of several hundred miles per hour will not make sharp bends.

SUMMARY

Tornadoes touch down as the circulation within the thunderstorm develops sufficiently to support them. The mesocyclone increases in strength as the surrounding air passes around the southern side of the thunderstorm, lightning heats the inner walls of the vortex, and rain is thrown outward by the centrifugal force of the circulation.

Tornadoes that develop in the center of the mesocyclone last longer and produce straight damage paths from faster moving thunderstorms and looping damage paths from slow moving thunderstorms. Tornadoes on the southern edge of the mesocyclone are carried by the thunderstorm rotation around to the front of the thunderstorm and then into the rain area where they dissipate. Such a pattern is frequently repeated with the formation of several different funnels.

The typical tornado damage path is 150 yards wide and 5 mi. long, with a forward speed of travel of 30 mph. Accurate measurements are not available to tell us the amount of pressure decrease in the tornado funnel or the speed of the winds. Tornadoes are more frequent in the Central United States, but more people are killed by tornadoes in the Southern United States. A single tornado outbreak in 1974 killed 315 people and the outbreak in 2011 killed 324.

Tornado watches are issued by the Storm Prediction Center when the atmospheric conditions are right for tornado development. Tornado warnings are issued when a direct threat of tornadoes exists. Warnings are based on the observation of a hook echo or tornado signature on the radar screen or eyewitness reports of tornadoes on the ground, and are issued by local National Weather Service offices and are transmitted to the public by radio and television stations. Individual warnings may be provided by cell phones, sirens or special weather radios.

The safest location to seek shelter from a tornado is in an underground reinforced concrete room or shelter. Concrete basements are safer than houses without basements. Northeast rooms in the basement and first floor are safer than other rooms for a typical tornado that approaches from the southwest.

7

Laboratory Tornadoes

As I pulled on my leather gloves and carefully placed the slices of dry ice to cover a large section of the floor, I felt a heightened sense of anticipation. If my calculations were correct the crosswinds more than three ft. above my head would interact perfectly with the rising airflow to duplicate the formation of one of nature's most fascinating phenomena. As I flipped the switches to start the air currents and flood lights, a hush fell over the group that had gathered to witness the experiment. For one long minute nothing happened beneath the simulator. As we watched the floor, a couple of yards from our feet, the dry ice cloud showed a convergence of air into a peak that suddenly shot past our eyes and above our heads into the vortex generator. To the delight of everyone a tornadolike vortex was born and continued a twisting, swaying dance that seemed to be designed especially for the occasion.

The creation of room-sized tornado simulations has been a keen interest of others as evidenced by Universal Studios paying my way to Hollywood to bring my published book describing how to make a tornado in the laboratory. They then built the Twister Building in Orlando that housed a 50 ft. copy of my miniature tornado that was viewed my millions of people. Details are in my autobiography, *Name Your Price*. Ned Kahn of the Exploratorium in San Francisco also consulted with me to build a tornado there, where people could touch the tornado simulation. This exhibit has continued to be a major attraction for the Exploratorium for many decades. A team of eight Japanese came from Fuji TV in Tokyo to film it in my laboratory. My main interest in creating a realistic tornado simulation was its potential value for research into the nature of tornadoes.

ATMOSPHERIC VORTICES

Vortices form in the atmosphere every day. These vortices are swirling masses of air that can be as small as your finger or so large that they cover half the United States as described for the frontal cyclone. One particular atmospheric vortex, the tornado, is generated about 1200 times per year in the United States. A common characteristic of all vortices is that they are composed of air generally rotating horizontally about a vertical axis. The vertical core can be composed of an updraft, downdraft, or it can contain no atmospheric motion.

The largest atmospheric vortices, other than the polar vortex, are the traveling frontal cyclones that represent a dominant feature of the weather of all midlatitude locations. One of the smaller visible atmospheric vortices is a **whirlwind** that can be seen on any hot summer afternoon in desert regions, and less frequently in other locations. While the whirlwind and hurricane are surface generated atmospheric vortices, the large midlatitude cyclone and the tornado are generated by air currents above the surface and spiral their way downward.

It is recognized that the large frontal cyclones are originated by the jetstream and extend down to the surface where they determine our weather on a particular day. It is now also recognized that tornadoes are generated within the interior air currents inside a thunderstorm several minutes before the circulation becomes strong enough to reach the ground as a tornado.

The difficulties of obtaining measurements within a tornado and within the generating region of a thunderstorm are obvious because of the high winds associated with a tornado. The use of dual Doppler radar has enabled us to see some of the air currents within a thunderstorm as it generates a tornado; however, radar is only sensitive to water droplets and does not see the air itself. Therefore, the winds of the tornado are still largely unmeasured, but it may be possible to model their development in the laboratory.

The nature of the atmosphere makes it difficult to analyze atmospheric storms in the laboratory as a chemist would analyze a solution in a test tube. But it is possible to generate a vortex in the laboratory and to investigate how a miniature tornado is related to atmospheric vortices formed outside.

GENERATING MINIATURE TORNADOES

A laboratory vortex that was created by placing a pan of heated water at the base inside a plexiglass cylinder with a circular opening at the top is shown in Figure 7-1. This was 8 ft. tall and had a fan at the top to draw the air out to create a vortex inside the cylinder. The cylinder had open slots on opposite sides to allow air to come in. This allowed the incoming air to start a spiraling flow as the air rose above the heated water to go out through the top; thus, an atmospheric vortex was created. This display of a vortex inside an 8 ft. cylinder did create a lot of interest at the Kansas State Fair in 1980 but the relationship between this vortex and a tornado or hurricane is very questionable.

Figure 7-1. This laboratory vortex produced in a cylinder 8 ft. tall and 2.5 ft. wide, designed by the author for the 1980 Kansas State Fair, was effective in generating considerable interest (Photograph by Greg Eagleman.)

An unconfined laboratory tornado was generated at the University of Kansas, with the assistance of Vincent Muirhead, by simulating the larger circulation of the mesocyclone vortex with a rotating cage and dynamic updraft in the center by creating a negative pressure with large vacuum tanks. The cage of 18 in. diameter was rotated at varying speeds while the pressure at the 2 in. diameter opening at the top center of the cage was varied from atmospheric pressure down to 0.1 atm. Different combinations of rotation and negative pressure produced tornadolike vortices of different appearance (Figure 7-2). More rotation produced a larger vortex with more irregular edges while more suction and less rotation produced a smaller vortex with smoother outer edges and a well-defined central core. This first tornado simulation was good for getting some measurements but a better match with outside atmospheric conditions was sought.

SIMULATING WINDS IN THUNDERSTORMS

Figure 7-2. An unconfined laboratory vortex can be generated by a combination of rotating air produced by a rotating screen with a suction up through to the center of rotation.

Immediately prior to the development of dual Doppler radar, a theoretical model, the double vortex thunderstorm model, was developed as described in a previous chapter, from mathematical equations to explain the winds required in a large thunderstorm to generate a tornado. Thunderstorms represent a barrier to the surrounding airflow and the application of the equations describing airflow around a barrier reveal that a double vortex would be generated within the thunderstorm to resist the surrounding airflow. The double vortex consists of counterclockwise circulation around a vertical axis in a large part of the southern part of a thunderstorm and clockwise rotation around a vertical axis in the northern part of a thunderstorm moving toward the northeast.

Tornadoes are known to originate in the southern part of a thunderstorm as shown previously in Figure 6-1. The counter rotating vortices within the thunderstorm are intensified by the crosswinds of the atmosphere composed of southwestly winds striking it in the upper and middle parts of a thunderstorm and southerly surrounding winds in the lower part of a thunderstorm moving toward the northeast. The moving thunderstorm increases the inflow of low-level air from the east and intensifies the opposing nature of the air currents.

Dual Doppler radar data has verified the existence of a double vortex in severe thunderstorms. Thus, the structure of a tornado producing thunderstorm is very organized with an updraft and crosswinds in the southern part of the thunderstorm that generate the tornado. This mechanism could be called a mechanical means for formation of tornadoes.

Such information on the wind structure within a thunderstorm that produces a tornado can be used to simulate these same opposing air currents in a laboratory to produce a laboratory tornado.

Figure 7-3. *An air current blowing around a curved metal shield will produce a vortex if there is an updraft at the top of the shield.*

LABORATORY TORNADOES PRODUCED BY BLOCKAGE

Since the flow of air around a barrier can be expected to produce vortices behind the blocked air current, it should be possible to combine this circulation around a barrier with an updraft to produce a laboratory tornado. Such a vortex was generated, by mounting above an opening in the ceiling, a 3/4 hp electric motor with a four-bladed propeller operating at 1,725 rpm. This produced an updraft, and ordinary 20 in. window fans were mounted below the ceiling to simulate the horizontal winds around a curved metal shield that formed a barrier to airflow. These 20 in. fans were 9 ft. above the floor. This arrangement produced a vortex as shown in Figure 7-3. The laboratory tornado extends down to a table located 5 ft. below the curved metal shield in this photograph, or down to the laboratory floor. The vortex is made visible by placing a layer of dry ice over the table or floor. The central core of the vortex is visible in Figure 7-3 and has an appearance very similar to some photographs of tornadoes and water spouts. Water spouts more frequently show the central core since they are generated by thunderstorms over

Figure 7-4 *Waterspouts frequently show the inner core because of lack of debris. (NOAA Photograph).*

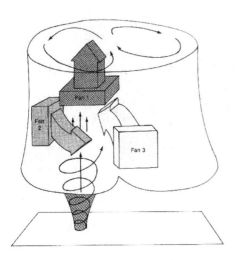

Figure 7-5. The laboratory tornado is generated by simulating the updraft within a thunderstorm and the air currents for a thunderstorm with a tornado in the southern part.

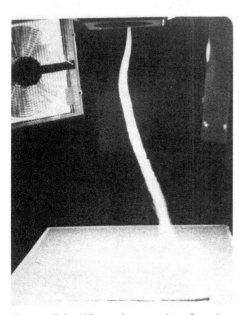

Figure 7-6. When the opening for the updraft is decreased, a vortex is still produced but with smaller diameter.

oceans or lakes where dust and debris does not hide the vortex structure as in Figure 7-4. This water spout formed near the Florida Keys and clearly shows the central core with rotating winds around it.

LABORATORY TORNADOES FROM CROSSWINDS

Since the thunderstorm is able to generate a tornado by using horizontal crosswinds combined with an updraft as shown in Figure 7-5, it should be possible to generate a tornado-like vortex in the laboratory by these same air currents. The two smaller fans were aimed in opposite directions to simulate the opposing horizontal winds, while the third fan simulated the thunderstorm updraft. The updraft was narrowed through a circular opening of 8 in. diameter. A photograph of the laboratory tornado produced by this arrangement is shown in Figure 7-6 as it extends from near the ceiling down to a table about 1.5 m below.

It was found that larger and stronger laboratory tornadoes were generated by making the updraft opening large enough to correspond to the diameter of the updraft fan. Figure 7-7 shows a vortex generated by this arrangement extending from the ceiling to the floor. The opening of the updraft area was 22 in. with the two 20 in. fans supplying crosscurrents to simulate the horizontal winds in the tornado generation region of the thunderstorm as shown in Figure 7-5. You can see a video of the miniature tornado in action on YouTube at: https://youtu.be/uV-MboO_fr4

Figure 7-7. The diameter of the full-length laboratory vortex can be decreased in particular regions by interfering slightly with the airflow surrounding it.

VARIATIONS IN CROSS-WINDS

A number of interesting variations in the direction and speed of the crosswinds were tested with the laboratory tornado. Figure 7-8 shows diagrams for some of the many variations in air currents that were tested. Diagram A in the figure corresponds to the photograph in Figure 7-3. The photograph in Figure 7-6 corresponds to diagram B. Diagram C shows a combination of crosswinds and updraft that might be expected to create a very strong vortex as both side fans are pointed toward the vortex and the updraft is pulling air upward. But it produces a weak intermittent vortex. The opposing crosscurrents in a horizontal direction feed the rotating air in the upper part of the vortex. The vortex forms, disappears, and then reforms as the three air currents interact in different ways.

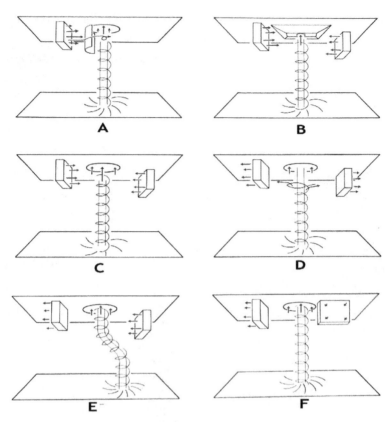

Figure 7-8. Different air currents used to create a tornadolike vortex in the author's laboratory are shown here.

Another combination of crosswinds that might be expected to produce a strong vortex is that of diagram D. It also produces an intermittent vortex. In this case the horizontal crosswinds are aimed outward from the updraft area in an effort to pull the outer rotating part of the vortex upward instead of feeding it as in previous diagram C. The vortex produced by these air currents is intermittent since the vortex is destroyed as the outer shell is pulled away from the central core. This rapidly reduces the strength of the vortex much like the effect of figure skaters extending their arms outward to slow their rate of spin. Thus, the laboratory tornado repeatedly forms and is destroyed by this effect or by attaching to a side fan (Figure 7-9).

Another interesting arrangement of air current was produced by removing one of the horizontal air currents shown in diagram D. A vortex does form but it is not sustained. The vortex was generated by the updraft and a single horizontal air current blowing away from the updraft area. This particular arrangement is interesting since it shows that the main requirement for generating an atmospheric vortex is a deficiency of air. As the air rushes in to fill the partial vacuum and begins a spiraling pattern the rate of air-flow into the area of lower pressure is reduced and this allows the whole process of filling the partial vacuum to be prolonged but in this case, it does not have the configuration to sustain the vortex.

One of the largest laboratory tornado-like vortices is generated by the somewhat surprising arrangement of air currents shown in diagram E. This laboratory tornado is more stable and not as intermittent as those produced by the previous arrangements of air currents. It is generated by the updraft with both the horizontal air currents going in the same direction. It is not obvious why this combination of air currents should create a stronger vortex than the arrangement of air currents previously shown, but it indicates that wind shear is an important parameter.

A steady and persistent laboratory tornado is also produced by the arrangement of air currents shown in diagram F, where the cross-currents blowing on different sides of the updraft are at a 90° angle to each other. This creates a strong vortex between the ceiling and the floor of the laboratory. It is interesting to note that these air currents are most similar to the winds in the atmosphere during the

Figure 7-9. After its generation, the vortex may attach to a side fan instead of the central updraft. This causes it to decay rapidly as indicated by the lower part of the vortex.

time of tornado formation, with easterly relative winds going into the lower central part of the thunderstorm as the winds on the back side of the thunderstorm come from a southwesterly direction.

OTHER OBSERVATIONS OF LABORATORY TORNADOES

A very interesting characteristic of the laboratory vortex is shown in Figure 7-9. After the vortex has been generated by an updraft and horizontal air currents, the vortex may attach itself to one of the side fans. It soon "winds down" in this position, however, since it has lost the source of energy for the vortex. This characteristic probably applies

Figure 7-10. The various shells of rotation can be seen in the lower part of this laboratory vortex. The central core and smooth vortex are surrounded by a shell with more irregular circulation

Understanding Severe and Unusual Weather

to other atmospheric vortices, such as tornadoes, and indicates that they would attach themselves to the nearest region with a net outflow of air.

The vortex is composed of several shells of circulation as shown in Figure 7-10. The outer shell can be stripped away leaving the central stronger circulation around the core itself. A destruction of this outer shell leads to more rapid decay of the whole vortex. The last stage of a tornado is frequently a rope stage as shown in Figure 7-11. This type of vortex occurs in the laboratory when the updraft is turned off. This may mean that such factors as interaction between

Figure 7-11. Only the inner core and smooth surrounding shell are visible in this rope-shaped laboratory tornado.

the tornado and the rain area, or improper air currents within the thunderstorm have been responsible for reducing the updraft support for the tornado and contribute to its decay.

Surface interactions can be investigated with laboratory vortices as shown in Figure 7-12. Some scientists in the past have suggested a downdraft within the core of a tornado while others have felt that an updraft should exist there. Direct measurements in a tornado are, of course, not available. But the confined laboratory tornado clearly shows that foam peaks upward into the core. This is consistent with lower pressure and an updraft through the core of the vortex.

If a traveling tornado is simulated by moving a sheet of plywood beneath the unconfined laboratory tornado it reveals airflow patterns that compliment and amplify tornado damage investigations. The path of the vortex across a viscous liquid coating on the plywood reveals airflow primarily in the direction of travel of the vortex (Figure 7-13). Slow movement produces a looping path while a faster rate of tornado travel produces a straight path. The detail of the inflow into the vortex as shown in Figure 7-13 is interesting in that it shows a major part of the inflow is in the direction of travel of the vortex but it has a component related to the circulating air in the vortex. Observations of damage patterns following a tornado show some of these same characteristics. The particular inflow characteristics at the base of the vortex are important as this determines how houses are damaged by a tornado.

An interesting observation of the laboratory vortex is that after

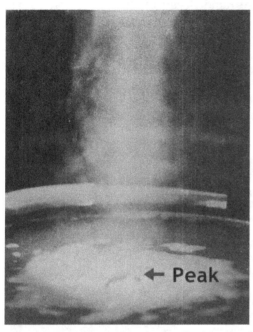

Figure 7-12. The inverted barometer effect is produced by the confined laboratory tornado. Low pressure inside the core of the vortex causes foam to form a peak that rises upward until it collapses or is broken and centrifuged out of the vortex

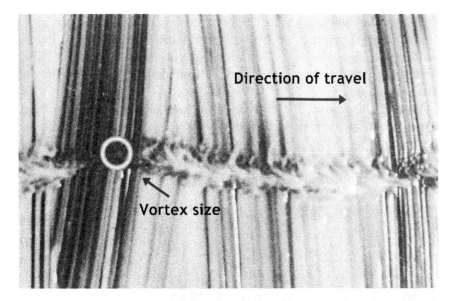

Figure 7-13. For a moving vortex, the major airflow within the base of the vortex was in the same direction as vortex travel.

it has been generated, it exists because of its own momentum for some time after the generation mechanism has been shut down as was shown in Figure 7-11. This illustrates that the air has momentum that allows it to maintain its velocity after it has been developed. Thus, the tornado may last for a short time after the generation mechanism has been eliminated.

APPLICATIONS

Since the laboratory vortex was generated by simulating some of the air currents within thunderstorms as developed theoretically and measured by dual Doppler radar, it is helpful in understanding some of the characteristics of tornadoes. One of the conclusions of these experiments with a laboratory vortex is that tornadoes may be generated by a variety of air currents within the atmosphere. This is undoubtedly one of the reasons why the smaller tornadoes are so difficult to forecast. If they can be generated by such a variety of crosswinds then the only hope in being able to forecast small tornadoes is to obtain information on the intensity of the thunderstorm updraft.

The updraft must be large enough to create a deficiency of air that will support a vortex from the central part or lower part of the cloud all the way down to the ground. Larger tornadoes are more easily forecast as they correspond to a more organized structure within the thunderstorm that consists of large counterrotating vortices. Such thunderstorms produce tornadoes that last longer and are more damaging.

One of the reasons for studying tornadoes in detail is to determine their particular formation mechanism and structure, since such an understanding may make it possible to consider modifying or controlling them sometime in the future. However, we have no operational programs attempting efforts in this direction based upon our current level of understanding of these atmospheric storms.

EXPLORATORIUM TORNADO

Figure 7-14 Tornado display at the Exploratorium continues to attract attention. Photo credit: Esther Kutnick, © Exploratorium, www.exploratorium.edu

The laboratory tornado received a lot of attention from newspapers in Lawrence, Topeka and Kansas City. It was also videoed by television stations in Kansas City, Topeka and Tokyo, Japan. It was covered in an article in the Preprint Volume of the 1971 National Conference on Severe Storms along with the double vortex thunderstorm model and presented to the audience at the conference. By some of these means it came to the attention of Ned Kahn at the Exploratorium in San Francisco in the early 1980's. He called me and asked if I could help him create a similar but larger tornado-like display. I agreed and gave him one of my books that described how I had created it in my laboratory. Over the next several months he called regularly as he struggled with the details of the display until it finally was completed. It has survived several transformations but continues to be of interest to many people and is still today one of the major attractions for the Exploratorium.

SUMMARY

Atmospheric vortices can be generated in the laboratory to simulate some of those that occur in the atmosphere. In general, vortices that are enclosed within cylinders or cages are not as realistic as unconfined vortices. One of the most realistic tornado-like vortices is generated by simulating the winds in a thunderstorm. These include opposing air currents in the presence of a large updraft. It can be shown in the laboratory that flow of air around a barrier will produce vortices behind the blocked air current.

Various combinations of crosswinds can be investigated for producing laboratory vortices. Several different combinations of cross-winds will interact with an updraft to produce a vortex. A persistent laboratory vortex extending from the ceiling of the laboratory to the floor was produced by air currents blown horizontally at an angle of 90° to each other; in combination with an updraft these produced a strong laboratory vortex. Laboratory vortices can be used to study the detail structure of a vortex, including its various component parts such as the updraft core and shells of circulation with various speeds around this core.

In addition to the generation mechanism, a laboratory vortex can be used to study the air currents as the vortex interacts with the surface beneath it. Such experiments show the major wind at the base of the vortex is in the same direction as that of the vortex travel with a smaller cross wind component.

You can see a video of my miniature tornado in action on YouTube at: https://youtu.be/LBzhAYuDnm8

Joe R. Eagleman

8

Softening the Blow

There was very little warning that this thunderstorm was any different from the hundreds of others that had passed over our house. At first, I dismissed the rising noise level outside as a low flying jet, but realized that a tornado was upon us as I saw the neighbors roof lift, spin, and suddenly disappear. As we tried to rush to the basement the windows shattered and we hit the floor. The next thing we knew we were suspended in air and the floor was ripped from beneath us. We were thrown against some bushes where the northeast part of our house had been. Our only thoughts were for our own safety and not of the magnitude of our loss at that moment.

DISTRIBUTION OF TORNADO DEATHS

The distribution of tornadoes is such that more tornadoes occur in the Central United States than any other area, comprising a band of greatest activity through Oklahoma, Kansas and Missouri as previously described. If tornado activity is combined with the number of people, the potential for tornado deaths should result. This produces a band of high risk in the Central United States because of many tornadoes and in the eastern United States because of the high density of people. However, observed tornado deaths are more concentrated in the southern United States, centering in Mississippi,

Alabama, Arkansas, Louisiana, and Texas. The design and type of houses, lack of basements, poor information concerning protection during tornadoes, and less concern about tornadoes may account for the higher number of deaths in these areas. Since it is impossible to control tornadoes it is important to have good information on designing houses and seeking protection during tornadoes, based on past observed effects of tornadoes.

This chapter contains the original classic studies of safest location in houses that shattered the advice that the Weather Service had heavily promoted for many years to seek shelter from a tornado in the southwest corner of any building.

The first study was of the Topeka tornado of 1966. This was a massive F5 (EF5 now) tornado that killed 17 people, damaged hundreds of houses and caused damages estimated at 1.8 billion dollars in todays dollars. This investigation showed that the southwest corner was the worst place for shelter from the storm.

The second study is of the F5 Lubbock tornado in 1970 that killed 26 people and caused almost a billion dollars damage in todays dollars. These two Kansas and Texas tornadoes remain two of the worst tornadoes in the history of each state.

Figure 8-1. The Topeka Tornado on June 8, 1966, was quite large, about 0.5 mi. in diameter at the ground. (Photograph by Perry Riddle)

TOPEKA TORNADO DAMAGE

The F5 Topeka tornado was very large, measuring about one half mile in diameter at the ground (Figure 8-1). It produced almost complete destruction in its path. Destruction of houses was produced by high wind speeds, related pressure changes, and debris. Boards, tree limbs, roofs, and other debris were carried in winds with speeds of over 260 mph. Such debris punctured the south and west walls of buildings. Looking to the southwest from Topeka is Burnett's Mound which was supposed to protect the city, according to Indian legends (Figure 8-2).

After the tornado came directly over Burnett's Mound it hit a residential section in Topeka composed of houses with walkout basements that all faced the southwest. The northeast side of the walkout basements were mostly underground, extending only about 3 ft. above ground. The walkout side on the southwest was more exposed to tornadic winds. The southwest rooms, in general, were destroyed, but the floor covering the poured concrete basements stayed in place in a high percentage of the cases (Figures 8-2 and 8-3). This was one reason why more people were not killed in this section.

A number of "survival" stories circulated from this section of the city after the tornado, including the children who hid under a pool table in the southwest rooms, the two couples who hid in a bathtub

Figure 8-2. Burnett's Mound, to the southwest of Topeka is shown in the background behind houses severely damaged by the tornado.

Figure 8-3. Walk-out basements such as this one generally had a part of the floor remaining over the northeastern section that is shown here.

with a mattress pulled over them on the main floor above the basement, and a party of people with one brave but stubborn woman who insisted on staying upstairs while the tornado passed. She was found hanging over a wall remnant, unhurt, after the tornado. Figure 8-4 shows the southwest side of one of these houses that was penetrated by debris and demolished by the high winds.

Observation of damaged houses showed that the larger rooms were less safe than the smaller rooms. One of the reasons for this is the

Figure 8-4. The southwest sides of the walk-out basements such as this one in Topeka were severely damaged by the tornado.

Figure 8-5. The south side of this multi-apartment house shows the amount of debris that bombarded and piled up against it.

added strength from the additional walls per unit area. The closets of a room are frequently left standing while the walls of the larger rooms are blown away by the tornado.

The next section of houses in the path of the tornado was composed of multiple family brick houses (Figure 8-5). These were apartment houses holding four families. Each house has a basement, mostly underground, and two stories above. The top story was the most unsafe while the basements were safest. Basements were unsafe only along the south sides where debris came through parts of the above-ground walls and through the windows.

Figure 8-6. Houses located on the southwestern corner of a block were frequently lifted and jammed against each other toward the north, as shown in this photograph. The roof in the middle belongs to the steps on the right.

Figure 8-7. McVicar Chapel on the Washburn University campus where 50 people sought shelter from the tornado.

Hundreds of single-family brick and wood frame houses were damaged or destroyed by the tornado. The brick walls of the houses were crumbled by the high winds in a manner similar to the frame houses. On the southwest corner of several blocks, houses were shoved to the north by the tornado coming from the southwest and piled together down the street (Figure 8-6). This fits the inflow wind pattern for the laboratory vortex that was previously shown in Figure 7-13.

Washburn University campus was badly damaged by the tornado (Figure 8-7). Most of the buildings were built from stones held in place by mortar, and these were badly crumbled by high winds. Most of the buildings had to be torn down and rebuilt as a result of the damage. McVicar Chapel held more than 50 people who were attending a musical recital during the tornado. They rushed downstairs as the tornado approached and by mistake sought shelter in the southeast part of the basement. This turned out to be the only possible place in which they could have survived. The southwest room was packed with large stones from the upper parts of the walls as they came crashing through the wood floors to pile up past the level of the ceiling of the lower rooms.

The next residential section contained large two-story homes. Many had weak foundations which allowed the whole house to be blown toward the northeast. Some of the houses were turned upside

Figure 8-8. Older houses with weak foundations frequently were turned upside down as the one in the central part of this photograph. The house on the left has been shifted from its previous location up from the stairs.

down as shown in Figure 8-8. Along one street more than ten of these houses lost their south porch to the tornado. Some of them also lost roofs (Figure 8-9).

Figure 8-9. South porches were uniformly removed from many houses by the Topeka Tornado.

Even if the framework is retained, the interior of the buildings may be severely damaged by pressure effects. Pressure damage may be severe due to the lower pressure inside the tornado as it passes over a house. Such differential pressure can cause a building to virtually explode. The greater internal pressure in houses has the beneficial effect of sending walls of houses outward instead of inward on top of persons located inside. Only south or west walls were occasionally blown inward by the stronger winds blowing in the direction that the tornado traveled.

In the downtown area, the tornado was wide enough to hit a 10-story building and a water tower which were separated from each other by a few blocks. The water tower was undamaged but the office building was gutted. All the windows were broken, the shades battered, and much of the framework was ripped away. The reinforced concrete structure of the office building remained undamaged. The tornado passed close enough to the State Capitol Building to pull out one of its copper panels from the dome.

The next residential section also contained two-story houses with weak foundations. Houses were moved downwind, turned upside down, and generally suffered extensive damage. Houses with poured concrete foundations would be expected to retain the floor above, but many did not because of either no bolts in the concrete or no nuts on the bolts. Houses with brick or concrete block foundations were

Figure 8-10. This house was pushed toward the northeast by the Topeka Tornado allowing the southern part of the house to drop into the basement. Sixteen people sought shelter in the southwest corner of this basement.

shoved to the northeast a few feet as in Figure 8-10. Much debris entered in the south and west sides of the basement, whereas the house slid over the ground in the northeast section which resulted in much less danger to those who could have sought shelter there. The concrete block foundation in Figure 8-10 failed and the whole house was pushed to the north. Sixteen people were in the southwest corner of this house where concrete blocks, boards, and pipes flew in. Fortunately, no one was killed in this house. In the northern part of the house, no debris had fallen in.

Damage statistics collected from houses with basements damaged by the tornado in Topeka showed that the south and southwest parts were most dangerous with 50% of them unsafe, while only 25% were unsafe in the northeast part of the submerged basements (Figure 8-11). The reasons for the unsafe locations included weak foundations, movement of the house in the direction of the tornado, and debris corning through the south and west sides of the structure.

Analysis of the walkout basements oriented with the southwest wall toward the tornado revealed that only 18% of them were unsafe in the northern parts, but 88% were unsafe in the southern rooms.

Damage statistics for the first floors of houses damaged by the tornado showed that the most unsafe areas were along the southwest, south, and southeast. The safest locations in the houses were in the northern sections where only 30% were unsafe compared to 50% unsafe in the southern rooms.

In 1948, a Texas tornado virtually destroyed an amateur weather observer's house. The observer was hanging on to a bedpost looking

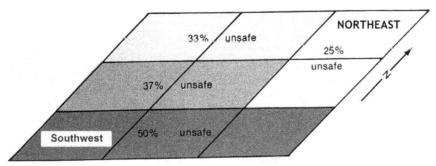

Figure 8-11. Damage statistics gathered after the Topeka Tornado showed that the most unsafe locations within houses were those along the south, and the safest locations were those opposite the approach direction of the tornado or the northeast part of basements.

Understanding Severe and Unusual Weather

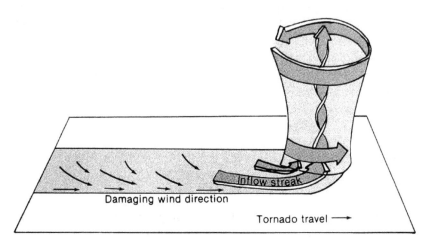

Figure 8-12. The most damaging winds are frequently in the direction that the tornado travels. This is because of the inflow of air from behind the vortex

up into the vortex. He described the funnel as being very smooth inside with brilliant lightning displays. He saw a house next door disintegrate the second the tornado touched it. Unfortunately, we do not have enough direct observations inside a tornado to understand them better. We must rely on field observations after a tornado or model them in the laboratory. Such tornado damage investigations and laboratory vortex observations indicate that much of the major damages are from winds going in the

Figure 8-13. Boards punctured the southern wall of this house. Such penetration by debris was common on the south and west walls and very uncommon on the east and north walls

129

direction of movement of the tornado or slightly to the right when looking in the direction the tornado traveled (Figure 8-12). This means that the inflow from the rear of the tornado provides the mass of air to be accelerated to velocities capable of causing damage. This also explains the observed debris punctures almost exclusively in the south and west sides of a house struck by the Topeka tornado coming from the southwest (Figure 8-13).

LUBBOCK TORNADO DAMAGE

The Lubbock tornado was the next single tornado to cause such extensive damage as the Topeka tornado. It formed from a very large thunderstorm just after sunset on May 11, 1970. The winds in this tornado were so strong that houses were completely destroyed and trees were uprooted or their bark was peeled off by the winds. Asphalt roofing from houses was forced into the dry ground so deep that it could not be pulled out (Figure 8-14). Metal roofing was wrapped around posts as if it were cloth. Large utility poles were ripped out or splintered. Much debris landed in a field east of Lubbock outside the damage area.

The tornado tore through a residential section containing some of

Figure 8-14. Asphalt shingles from the roofs of houses penetrated the soil with such force that they could not be pulled out after the Lubbock Tornado of May 11, 1970.

Figure 8-15. An indication of the extreme winds within the tornado can be gained from these photographs. These vehicles do not appear to have been rolled; the metal was simply battered and bent by the strong winds.

the finest houses in Lubbock. Basements are rarely built in this city or others in the South. Families sought shelter within their houses on the first floor or in some cases in outdoor storm shelters located in their backyards. It is amazing that the tornado killed only 26 people as it destroyed hundreds of houses. Boards were driven through walls into rooms inside, particularly on the south and west sides of houses. Metal was stripped away or bent on vehicles that apparently were not

Figure 8-16. Progressive damage is shown in this series of photographs. These range from the least damaged in the upper left to the most severely damaged in the lower right.

moved far (Figure 8-15). Houses with strong wall-to-concrete foundation connection were ripped off at a height of 3 ft. or so above the ground. In some cases the roof was lifted off and the walls fell outward. For some families, this tornado was the second or third that had destroyed their homes; most were rebuilding in the same location.

An aerial view shows how some of the roofs were damaged first at the peaks and eaves, then as a whole. Progressively greater damage is shown in Figure 8-16. Even in these well-constructed houses the roof-to-wall and wall-to-foundation connections show up as weak parts of a house.

Statistics collected on safest locations on the first floor of houses showed that 76% of them were unsafe in the southwest and south. The safest places were the central, north, and northeast rooms. The statistics

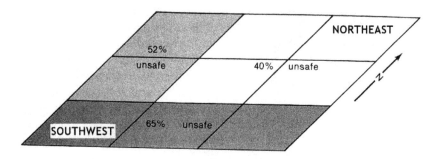

Figure 8-17. A composite of damage statistics gained from five different tornadoes has shown that the locations within houses that are most unsafe are those facing the approach direction of a tornado. Safer locations are opposite the approach direction of a tornado; since this is most often from the southwest, the northeastern part of a house is the safest.

for safest location in Lubbock were combined with similar investigations following four other tornadoes. These also show that the previously popularized southwest part of a house is the most unsafe during a tornado. Safest locations are those on the opposite side of the house from the approaching tornado. Since most tornadoes come from the southwest, those on the southwest are the most unsafe (Figure 8-17).

DESIGNING WIND RESISTANT HOUSES

The ability of houses to withstand high wind velocities can be greatly improved by proper planning and building design. More specifically, orientation and roof slope, quality of construction materials, and strength of connections of members greatly affect the strength of a house. Houses can be modeled and placed in a wind tunnel to obtain measurements during winds of various velocities. Models of houses were constructed and pressure measurements were obtained as the winds increased to more than 190 mph.

The angle of the roof was an important factor in the destruction of model houses by strong winds. Low angle roofs are lifted off from the same uplift forces that cause airplanes to fly. Air flowing over a 15° roof increases in speed with a resulting decrease in pressure that may be sufficient to lift the roof unless it is anchored firmly to the walls,

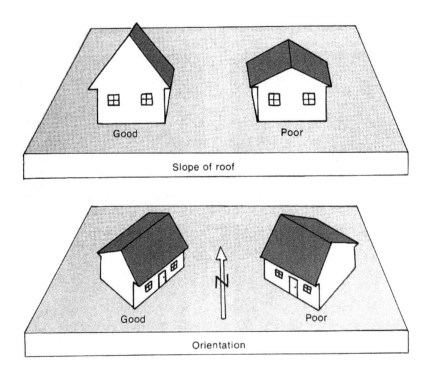

Figure 9-18. The design of a house contributes to its ability to withstand strong winds. Steeper slopes and houses with the short end oriented toward the southwest are more likely to withstand high winds.

This can be accomplished by more nails or by "hurricane clips," metal plates with teeth punched out to bite into wood after they are nailed in place. Another way to improve the ability of a roof to withstand winds is to use a steeper roof design. A roof with a 45° angle is not exposed to the reduced pressure from airflow over it since it is too steep for the "airplane wing effect." A roof with a 30° slope is much better at withstanding strong winds than one of lower angle (Figure 8-18). Every roof design has a critical wind angle, the orientation that produces the greatest pressure reduction over the roof. The 15° roof on a rectangular house has much greater pressure reduction, and hence more damage, if the winds are perpendicular to the long side. The best orientation for withstanding strong winds is with winds against the short side and parallel to the long side of the house. Thus turning the short side of a rectangular house with a low-angle roof toward the southwest would improve the chances for withstanding strong winds.

A flat roof is subjected to considerable negative pressure on the roof at the most upwind location. This develops as the airflow is directed upward over the edge of the roof, and turbulence is created there. The negative pressure thus created makes a flat roof more susceptible to wind damage than a 30 or 45° roof.

We also investigated the effects of other parts of a house, such as porches and chimneys, on wind damage. Models were prepared and tested with results that showed such modifications were helpful in reducing damage. The airflow was modified by the obstructions in a beneficial way for all roof angles.

Another test involved differences between one- and two-story houses. This showed only minor differences between the two designs if they had the same roof angles. For each design the two-story house was damaged sooner than the one-story house.

The patterns of failure were also of interest. The first parts of a house to be damaged as the winds increased were the eaves of the roof on the upwind side of the house and the downwind side of the peak of the roof. These structural damages began at about 150 mph. As part

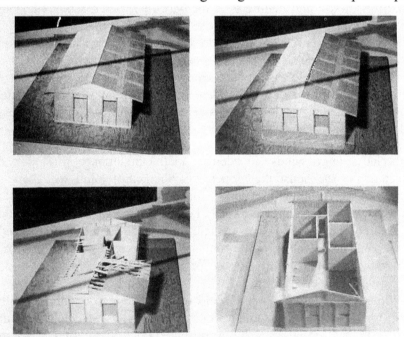

Figure 8-19. This series of photographs shows progressive damage during a destruction model test. Damage occurred first at the peak of the roof, then to the roof itself. Wind is blowing from left to right.

of the roof blew away, the potential for additional structural damage increased. A model building destruction sequence is shown in Figure 8-19.

In summary, wind damage is most likely to occur differentially according to roof angle and orientation. Steep roofs are safer in strong winds than low roof angles. Wind damage is most likely at the upwind edges and along the peaks, at the roof-wall joint or the wall-foundation joint. The downwind part of a house is likely to receive less damage than the upwind section.

The destruction model testing also showed that the southwest part of the first floor was unsafe because of the greater chance of structural damage in this part of the house, in addition to the greater chance for penetration of debris into this area. For houses with weak foundations, the southwest part is unsafe because of the tendency of tornadic winds to move the house off its foundation with displacement toward the northeast. Houses with good foundations (poured reinforced concrete) are also the most unsafe in the southwest part because of the increased risk of the house and ceiling over the basement being removed from over the southwest room before those in other locations.

Houses with vents were tested to determine the effect of ventilation on damage. In general, vents in locations comparable to doors and windows on a single story house caused it to receive greater damage if venting was provided on either the upwind or downwind side. However, vents on the downwind side near the peak of the roof allowed the house to withstand stronger winds.

A number of applications can be made that will allow houses to withstand stronger winds. In addition to the importance of slope and orientation of the house the proper strengthening of joints will allow a house to survive much stronger winds. Strengthening of the roof to walls by hurricane clips has already been mentioned. The walls should be securely anchored to the concrete foundation with bolts embedded in the concrete. Venting of the roof should be provided near the peak. This can serve the added purpose of releasing hot air from the attic during the summer.

While houses with poured reinforced concrete foundations with good wall to foundation connections are generally safe shelters from a tornado, additional security is provided by constructing a storm shelter as a part of the basement. Such an indoor shelter is preferable to an outdoor underground shelter because of the danger in getting to it during the short warning time for most tornadoes. The shelter should consist of concrete walls eight inches thick with a top slab of eight inches of concrete reinforced with metal bars connecting the walls and ceiling at nine inch intervals. A hallway with a 90° turn or strong door should form the entrance. This should prevent flying debris from entering the shelter. If such a shelter is constructed when the house is built it represents a very small additional investment.

SUMMARY

Since it is currently not possible to control or eliminate tornadoes, it is important to have proper information on seeking protection from tornadoes and on designing houses to withstand such high winds. Observations of tornado damaged houses have shown that smaller rooms are in general safer than larger rooms. In general, the southwest parts of houses are the most unsafe for a tornado coming from the southwest. Since this is the most common approach direction of tornadoes, the northeastern part of basements has been determined to be the safest from damage investigations. The reasons for this include the observed pattern of destruction where houses with weak foundations were moved several feet in the direction the tornado was going. This allowed debris and parts of the house to fall into the southwest part of the basement. With poured concrete reinforced with metal, the floor ordinarily remained over the basement. If part of it was removed by the tornado it more frequently was the southwestern part of the floor. For these reasons the northeastern part of basements is safest.

If basements are not available, the northeast part of the first floor is also generally the safest. The reason for this is that a large amount of debris is carried by high winds within a tornado. Such debris bombard the south and west walls sometimes penetrating through them, while no similar destruction of the east and north walls occurs. In addition, if any walls are blown inward, they are the south and west walls because of the strong winds blowing in the direction the tornado is going.

Houses can be designed to better withstand wind forces. The angle of the roof is an important factor in determining whether a house can withstand strong winds. A steep roof, with a 45° angle will withstand much higher wind speeds than a low angle roof of only 15° or a flat roof. The flat roof or low angle roof is subjected to strong lifting forces in the same way as an airplane wing.

The orientation of a house can improve its ability to withstand strong winds. The best orientation for a rectangular house with a low roof angle is with the short side turned toward the strongest winds; this means that the short side should be facing the southwest to improve the house's chance of withstanding the strong winds associated with a tornado. The orientation of houses with steeper roofs such as 45° is not so critical.

Houses can also be strengthened to withstand stronger winds by making the joints stronger at the roof-to-wall connections and the wall-to-foundation connection. This can be done with more bolts in the foundation or by using more nails at the other joints. Thus, it is possible to improve the ability of a house to withstand the strong winds associated with a tornado.

9

Fire from Above

At least we all made it to the 18th hole green before the rain. The rumbling from the darkened western sky had only vaguely registered as I tried to concentrate on the previous couple of shots. It was obviously going to rain as the growing thunderstorm changed the complexion of the afternoon sky drastically from its previous sunny state. We had been wet before while finishing a round of golf, and the extremely hot afternoon made the prospect almost inviting.

No one rushed the last round as the leading edge of the thunderstorm moved over us. We had heard thunder in the distance, but had no real sense of danger until the whole atmosphere seemed to be charged. My hair felt filled with electricity, and a glance at the others revealed the astonishing fact that their hair was literally standing on end. Just as I seemed to recall some advice for this situation the lightning discharge struck so close that I wondered if I was dead. When we recovered, we saw that the flag on the rod in the cup only 4 steps away was burned to a crisp by the lightning discharge. Lucky for us that we hadn't removed it yet, and that it was taller than us by the length of a hair.

HISTORICAL SETTING

Even small cumulonimbi are capable of generating thunder and lightning, giving rise to their common name of thunderstorms. The largest thunderstorms that produce tornadoes, rain, and hail also produce almost continuous lightning. Throughout history man has always had respect, awe, and even worship for lightning. He has not, however, always known how to deal with it. For example, in the eighteenth century it was the practice in Europe to ring church bells when thunderstorms developed, with the hope that this would somehow drive the lightning away. This practice was so dangerous to bell ringers that the people of Paris passed a law in 1786 making bell ringing during thunderstorms illegal.

The problem of living with the lightning hazard in the eighteenth century was amplified because churches were used for storing gunpowder. Most of the churches were designed with tall towers making them attractive targets for lightning discharges. Lightning and exploding gunpowder in churches were responsible for killing thousands of people in the eighteenth century. Since the nature of lightning was not understood at that time there was no information on where lightning would strike or why it struck particular locations.

It is interesting to note that some tall buildings were never touched by lightning even though they were constructed long before lightning was known to be caused by thunderstorms and electrical charges. One of these was the temple built by Solomon in Jerusalem. According to historians it experienced no damage from lightning over a ten century period. The wood and stone structure was completely covered with a thin layer of gold and had metal spikes on top of the temple protecting it from lightning damage.

STUDIES BY BENJAMIN FRANKLIN

The first studies of electricity were conducted around 1600 by William Gilbert, a physician for Queen Elizabeth I. He performed experiments that showed that material such as amber and glass would attract light bodies when rubbed. He gave this the name electrics taken from the Greek word electron, meaning amber. More than a century passed before much more progress in understanding electricity was made.

In 1746, Benjamin Franklin, at the age of 40 following successful business ventures, became interested in electricity. He discovered positive and negative charges and invented an electrical machine that could be cranked to generate sparks several inches long. He designed other experiments to gain more information on atmospheric electricity since very little was known about it at that time.

He devised an experiment to confirm that electricity was developed during thunderstorms. The experiment required a tall tower that could be insulated from the ground near its base. Franklin suggested that a spark could be produced across a gap between the insulated tower and

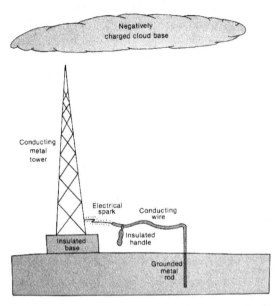

Figure 9-1. Experiment designed by Benjamin Franklin to test the charge buildup within thunderstorms. He suggested that a conducting wire be brought close to the insulated base of a metal tower while thunderstorm activity occurred overhead.

the ground by bringing a grounded wire near the metal base of the insulated tower (Figure 9-1). Although he had no tall tower in Philadelphia for testing it, he publicized this idea in Poor Richard's Almanac that was selling at the rate of 10,000 copies a year. The experiment was successfully demonstrated in Paris, France in 1752 when a spark was produced every time a thunderstorm came over the tower used for the experiment. Benjamin Franklin became famous in France and even before he knew the experiment had been successfully conducted, he was known as the man who could bring fire down from the heavens.

In the meantime, Benjamin Franklin was performing experiments with a kite. He was not satisfied just to know that thunderstorms contained electricity, but he tried to discover whether it was positively or negatively charged. In the spring of 1753, he found that all thunderstorms were negatively charged, but on June 6 of that year he measured positive charges during a thunderstorm. He concluded that nearly all thunderstorms have negatively charged bases, but occasionally one is positively charged. This remained the only reliable information on electricity in thunderstorms for 170 years. It wasn't until the 1920s that additional electrical experiments were performed to provide more information on the electrical nature of thunderstorms.

Benjamin Franklin invented the lightning rod with the first ones installed in 1756. A few years later in 1760, lightning struck a house in Philadelphia that was protected by a lightning rod and received no damage. Franklin had installed a lightning rod on his house, but it was not struck by lightning until 1787 when he was 81 years old. Benjamin Franklin's understanding of electricity was extraordinary for his time. He amused his friends by giving them electrical shocks during his kite experiments. Others such as King Louis XVI used a kite to shock 200 monks who were all holding hands. The electricity flowing through the kite string was also dangerous since Professor Reichmann in St. Petersburg was killed by ball lightning while performing experiments with a kite during a thunderstorm.

FORMS OF LIGHTNING

Most thunderstorms become negatively charged at their base and positively charged near the top through processes to be described later. The most common form of lightning (over 60%) is intracloud lightning, as discharges within a cloud occur between regions of positive and negative charges.

The next most common type of lightning is forked cloud to ground lightning. The lightning discharge is between the cloud base and the ground (Figure 9-2). The discharge develops as a negatively charged cloud base induces a positive charge on the surface of the earth by repelling other negative charges. Since opposite charges attract and similar charges repel each other, the negative charges at the thunderstorm base repel the negative charges within the surface layer of the earth. Therefore, the positive charge on the earth during a thunderstorm is an induced charge arising from the charge generation within the thunderstorm.

Forked lightning forms when the charge separation between the earth's surface and the base of the cloud develops to a critical level, just as electricity sparks across the gap in a spark plug, or from your finger to the door knob after walking over a nylon carpet. There are normally several individual lightning strokes during a single lightning

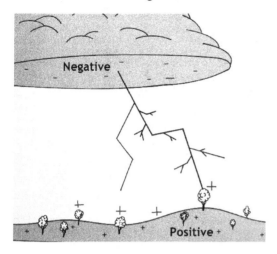

Figure 9-2. A common form of lightning is forked lightning that occurs as a discharge develops between the cloud base and the earth's surface.

flash. A flash of lightning lasts about 0.5 sec and is typically composed of three separate strokes. These occur so fast that you see a single image that appears to flicker. Sometimes as many as 25 strokes have been observed during a single lightning flash. Such a discharge lasts for more than a second and is, therefore, much more visible than the typical lightning flash.

Ribbon lightning sometimes forms if the winds are strong enough to displace the conducting channel beneath the cloud during the separate lightning strokes. A wind speed of

Figure 9-3 Forked lightning (NOAA photograph)

only 10 mph would displace the channel 14 ft. in one second. During a half second lightning discharge the multiple strokes through the same channel would be separated by a distance of 7 ft. between the location of the path of the first and last stroke. The diameter of the lightning discharge channel is only one or two inches. This channel conducts the charge from the cloud to the earth and from the earth back to the cloud during a typical lightning discharge. If winds displace the channel enough for it to be visible, the result is ribbon lightning.

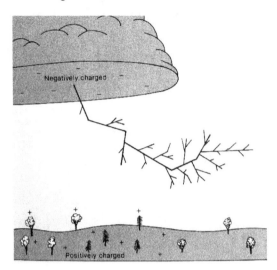

Figure 9-4. Air discharges occur when lightning is unable to complete a path from the negatively charged cloud base to the ground.

Sheet lightning is a type that occurs from distant thunderstorms as a flash beyond the horizon lights up the sky. Heat lightning is

Figure 9-5 Air discharge (Photograph by Micah Tindall)

Understanding Severe and Unusual Weather

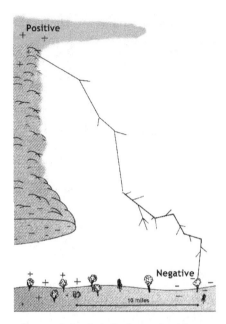

Figure 9-6. A bolt from the blue or superbolt occurs as the lightning travels for some distance horizontally before completing a path down to the ground.

similar to sheet lightning except that it occurs from thunderstorms located even further away so that no thunder is heard and the flashes are fainter.

Another form of lightning is an air discharge shown in Figure 9-4 and 9-5. This is a discharge into the air beneath the thunderstorm as the leader stroke (to be explained later) fails to reach the ground. This occurs frequently in desert regions where the cloud base is higher in the atmosphere.

A **bolt from the blue** or **superbolt** is a discharge from the upper positively charged part of a thunderstorm. The ground is normally slightly negative so the bolt must extend far enough away from the storm to get away from the induced positive charge under the storm. After traveling outward from the thunderstorm a few miles the leader reaches the ground producing a lightning strike (Figure 9-

Figure 9-7 Occasionally a superbolt lightning discharge occurs from the positively charged top of a thunderstorm to the negatively charged earth some distance away. Because of the greater distances, this lightning discharge may be many times stronger than an ordinary lightning discharge. (NOAA Photograph)

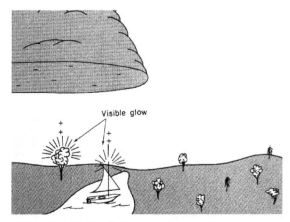

Figure 9-8. St. Elmo's Fire develops as the positively induced charge on tall objects on the earth develops sufficiently to cause an air glow.

6 and 9-7). The point that is struck may be as many as 10 miles away from the thunderstorm. The bolt from the blue is more dangerous because it is produced by a thunderstorm that is much farther away than normal and is also stronger.

St. Elmo's Fire is a form of electricity that is most likely to be seen at the peak of tall objects. These may have a bright glow because of the positive charges that remain as the negative charges drain away when a thunderstorm moves overhead (Figure 9-8). Towers, mountain peaks and other tall objects, even the horns of cattle, sometimes display this kind of electricity. Occasionally, the top of a thunderstorm glows with something like St. Elmo's Fire as the positive charges are concentrated there at extremely high levels.

Pilots of aircraft frequently observe St. Elmo's Fire if they must fly through thunderstorms. All commercial aircraft have radar on board to avoid thunderstorms, but this is not always possible. The lightning that sometimes strikes aircraft may be preceded by St. Elmo's Fire by five minutes or more. If lightning strikes an airplane it normally passes along the greatest distance, nose to tail or wingtip to wingtip, since the airplane is simply serving as a connection between the different parts of the cloud. If the plane is all metal it is a better conductor than the air for transferring the charge from one location to another. Metal aircraft are not usually damaged by a lightning strike. However, wood or fabric located between conducting metal may be ignited if struck by lightning. St. Elmo's fire was photographed (Figure 9-9) from NOAA's Hurricane Hunters plane on August 6, 2020.

Ball lightning is a form of lightning that is not well understood. An investigation of several hundred reports of ball lightning indicated that only three were possibly actual fire balls. The others could be accounted for by such explanations as images left on the eye after an intense lightning flash, since ball lightning usually occurs near ordinary lightning flashes.

Ball lightning consists of a ball of fire from 1 in. to a few feet in diameter. These may fall from the sky and explode or roll downhill until they strike an object and explode. They have an appearance resembling a soap bubble or a bubble of electricity. Ball lightning has come inside houses through windows or electrical outlets to float or roll across a room. One ball of fire about 2 ft. in diameter was reported to roll down a hall toward a person who stepped aside to let it pass. There are no verified photographs of ball lightning.

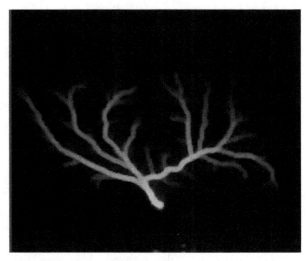

Figure 9-9 St. Elmo's fire probably on the wingtip photographed from NOAA's Hurricane Hunter plane.

When ordinary lightning strikes in deserts it leaves permanent evidence. The lightning flash melts some of the sand and forms a glassy channel about an inch in diameter and a yard or more in length. These glassy ribbons of glass, called **fulgurites**, have been used to study the climate of dry areas such as the Sahara Desert. The number and location of fulgurites is related to the number of thunderstorms and therefore gives some estimate of thunderstorm activity in the past.

DISTRIBUTION OF THUNDERSTORMS

Since lightning is developed by processes within thunderstorms the distribution of thunderstorms gives some information on the occurrence of lightning. In the United States, the average number of thunderstorms is greatest in the Florida Gulf Coast area where the mean number of days with thunderstorms is about 80 per year as previously shown in Figure 4-1. Another area that has a high frequency of thunderstorms is Colorado and New Mexico with thunderstorms on about 60 days during each year. The fewest number of thunderstorms occurs along the West Coast. In spite of the fact that the Pacific Northwest has the greatest mean annual rainfall in the United States, very few thunderstorms occur there. The rain in Washington and Oregon falls more as a continual mist.

Over the whole world about 44,000 thunderstorms are in action every day, while about 2,000 thunderstorms exist at any one time. These produce an average of about 100 lightning flashes per second. Even in the central Sahara Desert one thunderstorm per year normally occurs.

ORIGIN OF THUNDER

The thunder associated with a thunderstorm is produced by the discharge of electricity. Much of the energy of the lightning discharge is used in heating a channel of air to conduct the electricity from the thunderstorm to the ground. The conducting channel is heated to 50,000°F; about five times the temperature of the surface of the sun. As the air in this channel is heated very rapidly, in a few millionths of a second, it suddenly expands from a small fraction of an inch to an inch or two in diameter. This central conducting core

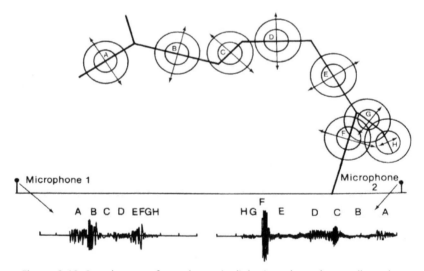

Figure 9-10. Sound waves from the main lightning channel as well as the branches combine to produce the overall pattern of claps and rumbles produced by a lightning discharge. The amplitude and duration of the sound wave produced by each of the individual elements is determined by the orientation of the segment and its distance from the observer. (After Scientific American, Vol. 233; 1, 1975.)

is surrounded by a glow discharge that may be several feet in diameter. The shock waves that result from the sudden expansion of air in the conducting channel spread outward as thunder.

The speed of sound is much slower than the speed of light. The sound wave travels only one mile in five seconds. This can be used to estimate the approximate distance to the lightning discharge by counting the time in seconds between the lightning flash and the sound of thunder, and dividing by 5 to obtain the distance in miles. If you count 10 seconds after the flash until you hear the thunder the storm is 2 miles away. The sound waves travel for only a limited distance so thunder will not be heard past this distance.

Rumbling occurs as discharges coming from different parts of the cloud arrive in short succession, with sounds also altered by echoes between clouds to produce a rumbling sound instead of the sharp crack produced by the main trunk of a nearby lightning discharge (Figure 9-10). The branches of the expanding air channel provide the crackling sound that accompanies many lightning flashes.

DEVELOPMENT OF ELECTRICITY IN CLOUDS

In the following section we will consider the details of just how lightning is generated by a thunderstorm and then in the next one the details of how a lightning stroke forms. A thunderstorm is able to generate lightning in about 20 minutes. Therefore, the charge separation mechanism within the thunderstorm must be rapid enough to operate within this time limit. A part of the explanation of charge separation involves the sudden freezing of liquid water droplets. Large updrafts can cause rapid freezing just above the freezing level in thunderstorms. It has been observed in the laboratory that a charge separation occurs if a temperature gradient exists in ice (Figure 9-11). Some of the molecules within ice are dissociated into positive and negative ions. Warmer temperatures cause greater dissociation. If a temperature gradient exists across ice it will result in the coldest part of the ice taking on a positive charge since positive ions are free to migrate from regions of high to low concentration while negative ions are not. This is called the **thermoelectric effect**.

As a supercooled water droplet freezes from the outside a temperature gradient may arise due to colder external temperatures. Small splinters of ice from the outer portion of the ice crystal are positively charged from ion migration, and are carried upward by updrafts to accumulate in the upper regions of the cloud, while the larger negatively charged ice particles fall to the lower part of the

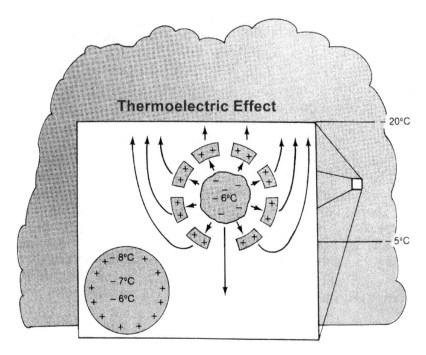

Figure 9-11. The charge separation in thunderstorms is related to the shattering of ice crystals as they are carried higher in the thunderstorm where the temperature is colder.

cloud. The region where this mechanism is effective is just above the freezing level, from 20°F to -25°F.

It has been well established that large updrafts and downdrafts exist in active thunderstorms. (This is another reason thunderstorms are avoided by pilots, in addition to the lightning and hail hazard.) The large updrafts further contribute to charge separation and the development of lightning within the required time of 20 minutes. After a slight buildup of negative charge at the base and positive charge at the top of a cloud by the shattering of freezing cloud droplets, the large ice pellets have an induced negative charge on their tops and positive charge on their bottoms. When this occurs the smaller cloud droplets within an updraft strike the lower surface of large ice crystals or pellets and bleed off some of their positive charge, leaving the small droplets positively charged and the large ice pellets negatively charged (Figure 9-12). The smaller droplets are then carried to the top of the thunderstorm by updrafts, adding to the

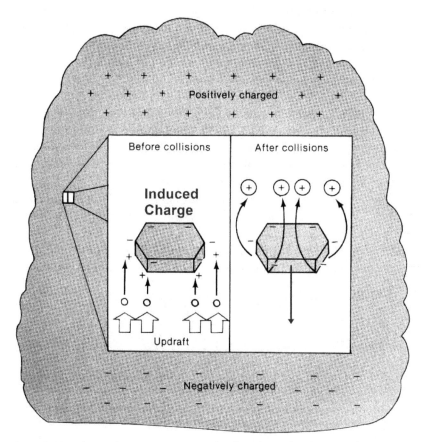

Figure 9-12. After a charge separation is developed in a thunderstorm because of the thermo- electric effect, ice crystals within the central part of a thunderstorm have a positive charge on their lower surface because of induction. There positive charges are bled off by updrafts containing smaller water droplets, thus intensifying the charge separation.

positive charge there, while the larger ice pellets with a more negative charge fall into the lower part of the cloud, adding to the negative charge there. Thus, the small cumulus cloud may grow into a tremendous electricity generator capable of developing almost continuous lightning discharges from the only resources available-air, water and dust particles.

It has been observed that a relationship exists between lightning discharges and rain on the ground. A major lightning flash often appears to be followed by a down rush of large raindrops at the ground. The electrical charges may help hold the cloud together as the positively charged cloud top pulls on all the negatively charged water drops at the base of the cloud. A sudden release of the electrical charges by lightning may allow the raindrops to fall more rapidly.

NATURE OF THE LIGHTNING FLASH

The lightning flash consists of several strokes. The first of these is a slower, stepped leader stroke. The stepped leader is slower because it must develop the conducting channel from the thunderstorm base to the ground. As the charge builds in the cloud and the charge separation reaches the sparking level it seeks the best channel to the highest or best conducting ground location producing forked paths since many of the channels do not lead anywhere. Eventually a path is developed from the cloud to the ground.

After the stepped leader develops a conducting channel to the ground the return streamer occurs (Figure 9-13). This is a discharge of

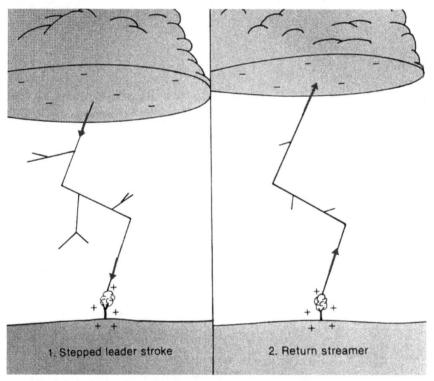

Figure 9-13. The stepped leader first builds a lightning channel down to the ground. This is followed by the return streamer as a discharge occurs back from the ground to the cloud.

electricity from the ground back to the cloud. The return streamer is a lot more brilliant than the leader stroke and is ordinarily the observed lightning stroke because of this brilliance. After a channel has been developed an average of three leader strokes and three return streamers occur through one specific conducting channel. The channel is made conductive for electricity by the extremely hot temperatures that develop ionized air molecules. Oxygen molecules, for example, are ionized into single oxygen atoms. Since these are charged, the electricity readily flows through the channel.

Lightning can be photographed by an ordinary camera at night by opening the lens when a thunderstorm is within view. A lightning discharge then produces an image on the film. An ordinary camera cannot usually obtain lightning photographs during the day since an open lens will overexpose the film before lightning occurs, but unusually long discharges can be photographed during the day by watching closely for them. The darkness from a severe thunderstorm may also allow lightning to be photographed (Figure 9-14).

The successive lightning strokes within a single flash discharge layers at increasing altitudes within the cloud. The first return streamer may neutralize the lowest layer of the thunderstorm with the next streamer neutralizing a slightly higher layer and so on. This allows more of the negative charges in the thunderstorm to be neutralized than would

Figure 9-14 Photograph of a wall cloud and lightning discharge. (NSSL Photograph)

otherwise be possible from a single stroke of lightning.

One of the beneficial effects of lightning is the production of nitrogen fertilizer for plants. As the air is heated by the leader stroke oxygen and nitrogen molecules are combined into a form usable by vegetation. This nitrogen fertilizer is carried by rain into the soil for plant use. Although the amount of fertilizer production in areas of heavy thunderstorm activity may amount to only 5 lbs./acre each year, it is interesting to note that it is applied at exactly the time when it is most useful to the growing plant. In comparison, 250 lbs./acre of manufactured nitrogen fertilizer applied to a crop in dry weather is of no value and may even "burn" the crop if the dry weather continues.

LIGHTNING PROTECTION

Lightning damages are substantial each year in the United States. According to insurance claims more than $900 million in lightning claims were paid out in 2019 to nearly 77,000 policy holders. This amount was also paid in the previous two years.

The best way of protecting houses from lightning is by use of lightning rods. Investigations of hundreds of fires have shown that 95% of the buildings that burned from lightning had no lightning rods and most of the remaining 5% with lightning rods had major defects in their methods of installation. The lightning rod is effective in protecting

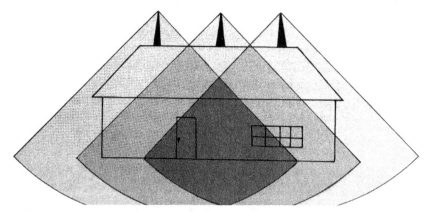

Figure 9-15. Lightning rods can be expected to protect an area enclosed within an angle 45° outward from the rod. The number of lightning rods required to protect a particular house is therefore determined by the height of the rods and the size of the house.

Figure 9-16. Lightning currents are evident in this photograph taken three days after a lightning strike on a golf course in Tucson. The five-foot fiberglass flag pole was tall enough to attract lightning. (Courtesy of E. Philip Krider).

buildings from damage only if it is properly grounded.

A lightning rod has a zone of protection beneath it in a cone of about 45° (Figure 9-15). A lightning rod consists of a metal rod extending above the highest point of the building to be protected with the other end of the rod extending to moist soil beneath the surface of the ground. Three feet or less beneath the surface is generally sufficient to anchor a rod into moist soil. Sharp bends in the lightning rod near other grounded routes that the lightning could take must be avoided. Lightning can jump across if sharp bends occur in the rod near metal pipes or other grounded conductors.

The spread of electricity from a lightning strike on a golf course can be seen in Figure 9-16 from the burned grass. The flag was obviously not well grounded in deep moist soil and the electricity was neutralized as it spread horizontally. This effect would be expected if only the first few inches of soil were moist with a dry layer beneath.

Taller objects are more likely to be struck by lightning. An object 300 ft. high located in a region where there are 30 days with thunderstorms per year can be expected to receive three lightning strikes per year. An object 900 ft. high could be expected to receive about ten strikes per year if located in an area with a similar frequency of thunderstorms. Radio, TV, and other towers are continually struck by lightning. Radio and television stations generally have auxiliary generators for this reason.

Lightning kills more people in the United States each year than most other weather events, although tornadoes and hailstorms cause much more property damage. For the 35 years ending in 1975 the average lightning deaths were 198 per year and the average deaths from tornadoes was 136. For the most recent 30-year period lightning deaths averaged only 43 per year. This significant improvement means people are better prepared to dealing with the threat of lightning.

Lightning deaths are normally singular events, but occasionally there are multiple deaths, such as occurred in December, 1963, when lightning struck a jet passenger plane over Elkton, Maryland, killing all 81 persons aboard. The second deadliest lightning strike occurred in July, 1961, near Clinton, North Carolina, when eight of nine people were killed as lightning struck a tobacco barn and passed through the metal heating system that the victims were leaning against. Another lightning tragedy occurred in August, 1960, when a young couple in Bay City, Michigan, were struck and killed by separate bolts of lightning while dashing across the street to their home. Lightning deaths are slightly more common in the eastern and central United States than in the western United States.

Lightning deaths occur more frequently to persons in open country than inside buildings. Most of the fatalities, 52 %, occur in the open, 38% in houses or barns, and 10% under trees (Figure 9-17). Unsafe locations inside houses are near metal water pipes, metal appliances, and telephones with above ground lines. Dangerous locations outside are on mountaintops and hilltops (or the highest ground in the area), under isolated trees, near wire fences and on open tractors with implements in the ground.

Safer locations in the open are in valleys or under a small tree that is located a few hundred yards away from larger trees. Automobiles are ordinarily safe during lightning storms since they surround a person with metal and if they are struck the electricity flows through the metal to the ground.

A lightning strike does not always kill a person. In fact, only about 10% of those struck are killed but most of the other 90% are left with injuries. One farmer who was plowing during a thunderstorm was struck and knocked unconscious. He was taken to a hospital where

Understanding Severe and Unusual Weather

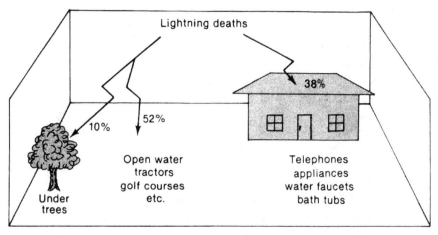

Figure 9-17. Unsafe locations during thunderstorms are under tall trees, in the open on open tractors and golf courses, or in houses near appliances or plumbing.

he survived, even though the lightning had burned a hole in his hat and had also burned a streak down his back. Recovery was similar to that following a stroke. I learned this from one of my students in my Unusual Weather class. This class filled the largest auditorium at KU holding 325 students every semester. The first time I asked if anyone had been struck by lightning, I didn't expect a positive answer. But I was wrong, because I can't recall a single semester when I didn't have one to three persons hold up their hand from out of 325 people. The hands went up timidly as would be an appropriate response to such an event. Many times, it was a friend or relative that had been struck. When I inquired about the circumstances the answers were reasonable, such as the tractor incident just described. A family member found the tractor going in circles and got the man to a hospital in time.

Lightning sometimes gives a warning a few seconds before it strikes. If your hair stands on end during a thunderstorm immediately squat as low as possible since this is an indication that you are about to be struck by lightning. Don't lay flat on the ground because the difference in potential between your head and feet may be enough to kill you if the strike is close to you.

It is important to know that it is possible to revive many people who have been struck by lightning and are apparently dead. A lightning strike is likely to stop the heart. If there is no other damage to vital organs the victim may be revived by immediate application of cardiopulmonary resuscitation. This should be continued until an ambulance arrives.

Some people have survived more than one lightning strike. The Guinness Book of World Records lists a man who has survived five lightning strikes. During his job as a park ranger in Virginia he lost his big toe nail in 1942 to lightning, his eye brows in 1969, had his shoulder seared in 1970, and his hair set on fire in 1972 and 1973.

Lightning strikes have been known to be beneficial. In 1782 a paralyzed member of the household of the Duke of Kent was struck by lightning. He was immediately cured of the paralysis.

More recently in 1980 lightning struck Edwin Robinson of Falmouth, Maine who was blind and deaf from a head injury suffered in 1971. His sight and hearing returned slowly within a few months. When in New York for an appearance on ABC-TV's "Good Morning America" he said his scalp "felt funny, like whiskers on my face." The lightning strike was also changing his baldness to a thick head of hair at age 62.

In addition to the effect of lightning on people and their property, lightning is responsible for starting many of the forest fires that burn every year in western states, although more than half of the fires are started by careless people. A good example of the forest fires were those in 2020. From January 1 to October 12 there were 45,635 wildfires compared with 42,821 wildfires in 2019, according to the National Interagency Fire Center. About 8.3 million acres were burned in 2020. On August 17 a series of lightning strikes started hundreds of fires across Northern California, dubbed the Lightning Complex fires. The largest was located in five counties in northern California near San Francisco and this was the third largest fire on record in the state. In California, fires were burning from the north all the way down to the Mexican border, stretching across approximately 800 miles of landscape.

Large fires were also burning at this time in Idaho, Montana, Oregon, Washington and five other states consuming over 2 million acres. In Oregon thousands of residents evacuated their homes to escape the flames that scorched more than 230,000 acres. In Washington, more acres had been burned in 2020 than in the past 12 fire seasons. The fires were being fueled by continuing dry conditions.

It has been suggested that better forest management could prevent the spread of many of these fires. If the forest floor is not kept clean the accumulated dead vegetation is easily ignited and serves as fodder for the continuation and spread of the fires.

GEOSTATIONARY LIGHTNING MAPPER

Since 2017 lightning flashes have been photographed from space. The Geostationary Lightning Mapper (GLM) is a satellite-borne single channel, optical instrument on the GOES-16 satellite. This allows measurements over the United States with continuous views capable of providing lightning detection at a rate never before obtained from space. All forms of lightning are recorded during both day and night, continuously, with a high spatial resolution.

The availability of lightning data increases severe storm warning lead time, gives earlier indication of impending lightning strikes to the ground, and total lightning coverage (Figure 9-18).

Scientists can now study electrical storms over dimensions ranging from

Figure 9-18. Lightning strikes on January 1, 2021 were in thunderstorms along the cold front of a frontal cyclone. This major ice and snow storm affected half the United States. The strikes are shown from the Global Lightning Mapper as small x's. They help identify developing severe thunderstorms.

the Earth's radius all the way down to individual thunderstorms. Disseminating lightning information in near real time, on a continuous basis with other observable data, such as radar returns, cloud images, and other meteorological variables provides invaluable data to aid weather forecasters in detecting severe storms in time to give advance warning to the public. The sudden increase in flash rate that has been found to be related to storm severity can now be detected for any region in the United States.

SUMMARY

Benjamin Franklin greatly advanced our knowledge of electricity. He investigated thunderstorm electricity with kite experiments and determined that most thunderstorms have large negative charges at their base. He also developed the lightning rod that is still used for the protection of buildings.

The most common form of lightning is forked lightning. Air discharges occur as the leader stroke beneath the cloud base fails to reach the ground. A bolt from the blue occurs if the air discharge travels for several miles and then reaches the ground farther away from the thunderstorm. St. Elmo's Fire may be seen as the tops of tall objects become sufficiently charged by a thunderstorm overhead to emit a continuous glow. Ball lightning is another form of lightning that develops as a ball falls from the cloud base and explodes as it strikes the ground, or forms near a lightning discharge and floats through the air.

Thunder is produced by lightning discharge as the conducting channel of air is heated to five times the temperature of the sun in a very short period of time. The sudden expansion of the air in the conducting channel creates shock waves that are heard as thunder.

The lightning flash consists of several strokes. The first of these is a stepped leader that creates the conducting channel from the cloud base to the ground. After the conducting channel is developed a return streamer occurs from the ground to the cloud. The hot air channel conducts electricity because the molecules are ionized and therefore charged. A side effect of the extremely high temperature is the creation of a small amount of nitrogen fertilizer.

The development of electricity in clouds is associated with the raindrop formation process along with updrafts and downdrafts within the thunderstorm. Spontaneous freezing and shattering of supercooled water droplets helps initiate the charge separation within a thunderstorm. As the top of the storm becomes positively charged and the base negatively charged an induced charge occurs on all of the ice particles and water droplets within the central part of the thunderstorm. The small cloud droplets within an intense updraft bleed off the negative charges from the lower side of larger ice pellets. The concentration in the upper part of the thunderstorm of smaller cloud droplets that are positively charged intensifies the charge separation process.

Lightning is responsible for over 40 deaths per year in the United States and about 900 million dollars' worth of damage per year. Properly installed lightning rods are effective in protecting buildings from lightning damage. Over one-half the lightning deaths occur outdoors with dangerous locations being near open water, on a tractor with an implement in the ground and no cabin, or on a golf course with a golf club over your head. Other unsafe locations are under large trees and indoors near telephones or metal appliances if telephone or electrical wires are above ground.

The damages from forest fires in the western states are substantial every year. Many of these are caused by lightning and are exaggerated by poor forest management.

Watch a video of lightning formation in a cloud on YouTube.
https://youtu.be/ixoUkqKnTNs

Joe R. Eagleman

10

Ice from the Sky

At first, I thought it was only a bad dream, but as I awoke more fully I realized that the pounding sounds on the roof were real. The continuous noises were punctuated by intermittent loud thumps. Occasional lightning flashes revealed the accumulating hailstones on the lawn. Most were golf ball size but scattered among them were chunks of ice as large as baseballs. One of them appeared to be larger than a softball!

The sound of breaking glass drew me to the kitchen where a large chunk of ice lay on the floor among pieces of the broken west window. The hailstone was larger than my closed fist and contained knobs all over it similar to my knuckles. After placing the hailstone on the hearth one blow from a hammer shattered it into several pieces that showed an interesting pattern of alternating layers of ice within it. My interest in this was dampened, however, since I did not need a vivid imagination to know what such hailstones were doing to my new car parked outside.

LOSSES FROM HAIL

In some parts of the United States hailstorms frequently damage personal property, such as automobiles and houses, in addition to the effect of large hailstones on plant and animal life. Three persons have been killed in the United States by large hail. A farmer near Lubbock, Texas, was caught outside on May 13, 1930, during a hailstorm and was beaten so severely by large hail that he died within a few hours. Softball size hail killed a baby in Fort Collins, Colorado, July 30, 1979. This storm injured 25 others and extensively damaged 2500 automobiles and 2000 houses. The third person killed by hailstones in the United States was a boater on Lake Worth, Texas, on March 29, 2000. Reports indicate that many people have been killed in India by hailstorms. A hailstorm in the Moradabad district of India killed 246 people on April 30, 1888. A 2017 paper in the journal, Weather, Climate, and Society, notes that this is the highest hailstorm-related mortality accepted by the World Meteorological Organization. Hailstones are also occasionally large enough to kill cows and horses; sixteen horses were killed in South Dakota on July 5, 1891, by very large hailstones.

Hail damages amount to several billion dollars per year. Damage to agricultural crops is extensive in the United States (Figure 10-1) and amounts to a billion dollars in some years. An average of about 2% of the nation's crop production is lost to hailstorms each year. In some areas of the Great Plains about 20% of the value of crops is lost to hail annually. Hail insurance is a major part of the cost of crop production.

Personal property damage is also a significant loss from hail storms (Figure 8-2). Large cities frequently suffer extensive damage from

Figure 10-1. Hail can completely destroy agricultural crops. A com crop such as that shown above would be unable to recover from the effects of a hailstorm.

Figure 10-2. Large hail is also a threat to various types of personal property including houses, automobiles, and aircraft. The severely battered wings of this airplane show the effects of flying through a thunderstorm that contained many hailstones. (Courtesy of NOAA.)

hailstorms. Over the last few years, the top five cities for average annual hail loss claims rounded to the nearest thousand were: Omaha, Neb. (18,000); Denver, Colo. (17,000); Colorado Springs, Colo. (12,000); McKinney, Texas (11,000); and Dallas, Texas (8,000).

OCCURRENCE OF HAILSTORMS

The distribution of hailstorms across the United States is slightly different from the distribution of thunderstorms and tornadoes. The area of maximum hail might be expected to be the same as the area of maximum thunderstorms, but this is not the case. The maximum occurrence of thunderstorms is in Florida, with a secondary maximum in the central United States in southern Colorado and New Mexico, while the maximum occurrence of hail is in northeastern Colorado and southeastern Wyoming (Figure10-3). As many as 8 days per year on the average have hailstorms in this location, while hail occurs an average of four days per year in the area encompassing parts of South Dakota, Nebraska, Kansas, Oklahoma,

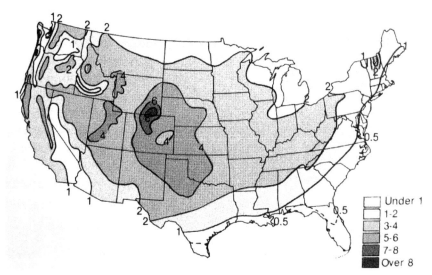

Figure 10-3. Hailstorms are most common in an area extending from Texas through South Dakota. They occur most often in Northern Colorado and Southeastern Wyoming. Because of the intensity of agricultural activities western Kansas suffers more damage to agricultural crops than other locations. (After Environmental Data and Information Services, NOAA.)

Texas, and New Mexico. Hail very rarely occurs in the Gulf Coast and Florida area, in comparison to the large number of thunderstorms there. This shows that specific conditions must exist in a thunderstorm to produce hail.

The season of major hailstorm occurrence is related to the location of the jetstream. Hailstorms start to become a problem in Texas and Oklahoma during April. By mid-May they are a greater threat northward over Kansas and Colorado, and during June hailstorms are more likely in Wyoming, South Dakota, and Montana. The region of major hailstorm activity corresponds to the migration of the jetstream. Hailstorms are more likely in the spring farther south with the band of greatest occurrence moving to the northern United States during the summer as the jetstream follows this pattern. Hailstorms are similar to tornadoes in this respect, since they also follow the movement of the jetstream.

Hailstorms over the whole United States are more abundant in May and June than other months of the year (Figure 10-4). They also frequently form in July and August making late spring and summer the time of major hailstorm activity. Since hailstones are composed of ice, they might be expected to be more common in the wintertime

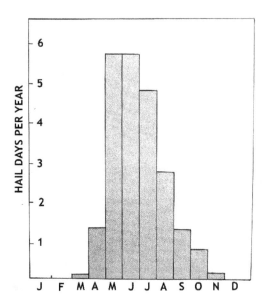

Figure 10-4. The monthly distribution of hailstorms shows they are more common in May and June with considerable activity also in July and August.

when it is colder, but this is not the case. Large cumulonimbus with high velocity updrafts are required to form large hailstones and these occur in the warm rather than the cold season. Thunderstorms develop more frequently as surface heating causes the air to become buoyant and rise to great heights resulting in hail formation. The most frequent time of hail is 3 to 4 pm with almost all hailstorms occurring between 2 and 6 pm. This emphasizes the importance of surface heating and the development of large thunderstorms in the atmosphere.

Hailstones have accumulated on the ground to depths of more than 1 ft. Hail fell at Seldon, Kansas, from 5:15 to 6:40 pm on June 3, 1959, with accumulations of marble-size hail to a depth of 18 in. Thunderstorms that produce heavy hail also frequently produce considerable rain. Runoff from the storm may accumulate hailstones into drifts that have grown to depths of more than 6 ft.

CHARACTERISTICS OF HAILSTONES

To be classified as **hail** it must fall from a convective storm and the particle size must be greater than or equal to 0.2 inches. The official definition of **severe hail** used by the National Weather Service is hail that is 1 inch or greater.

The appearance of a hailstone that is cut in half is similar to an onion, since it is composed of several layers (Figure10-5). The layers of a hailstone consist of alternating milky opaque ice and clear ice. The environment within the thunderstorm determines the appearance of

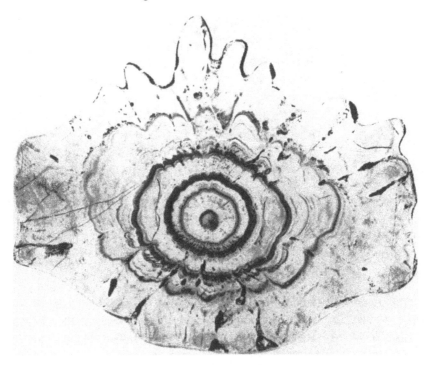

Figure 10-5. This photograph of the largest hailstone recorded from 1970 to 2010 shows six alternating layers made up of clear or milky appearing ice. This hailstone fell near Coffeyville, Kansas and measured over 7 1/2 in. in diameter. (Courtesy of NCAR.)

the ice within a hailstone. If the water freezes very slowly, bubbles have time to get out of the water producing a layer of clear ice. Rapid freezing traps the bubbles within the ice, producing an opaque milky layer. Rapid freezing occurs at cold temperatures high in the thunderstorm. Snowflakes and ice crystals are also likely to be present at this level. Some of these become entrapped within the outer layer as it forms on the hailstone contributing to the milky, opaque appearance of the ice formed high above the freezing level. Thus, the alternating rings in the hailstone correspond to the regions in the thunderstorm where the ice was formed above and below the freezing level. Small hailstones are usually round in shape, but large hailstones are seldom round and smooth. Large hailstones are often knobby and elongated. Sometimes large stones are conical shaped or flat. The most typical shape of a large hailstone is an oblate spheroid that has a very uneven and knobby outer layer.

Studies conducted in wind tunnels with ice suspended in a stream of high-speed air show some of the reasons for the different shapes of

Figure 10-6 Largest hailstone ever recorded in the United States fell in the yard of Lee Scott of Vivian, South Dakota, on July 23, 2010. (NCAR Photograph)

hailstones. The dynamics of airflow around chunks of ice help determine the various shapes. The optimum shape of a hailstone is related to its size and orientation within the updrafts with the dynamics of airflow dictating that a flat shape is more likely at one size and a conical shape at another. The rough outer layer of large hailstones develops as masses of water and ice accumulate and freeze because of the lower temperature of the ice as it falls from the very cold atmosphere above.

The size of hailstones ranges from 0.2" to 8". The record size for several decades (Figure 10-5) was the one that fell near Coffeyville, Kansas, on March 9, 1970, and measured 17" in circumference, 7.0" in diameter and weighed 1.7 lbs. The largest officially recognized hailstone on record to have been 'captured' in the U.S. was that which fell near Vivian, South Dakota in 2010 on July 23rd. It measured 8.0" in diameter, 18 ½" in circumference, and weighed in at 1.9375 pounds. Mr. Lee Scott, who collected, the monster stone originally planned to make daiquiris out of the hailstone but fortunately thought better and placed it in a freezer before turning it over to the National Weather Service for certification.

The size distribution of hail is slightly different in various geographical locations. Some studies have shown that the average hail size is 0.7" in Colorado while smaller hail is more common in New England. The most frequent size of hail is pea-size. Hailstones the size of grapes fall quite frequently, while walnut size hail falls less frequently. Still less often golf ball, tennis ball, baseball, and softball size hail is generated by thunderstorms. The eastern United States occasionally has large hail, but more often hailstones are small. This is related to the size and severity of the thunderstorm. Great Plains thunderstorms are generally larger and capable of producing the largest hailstones. The largest recorded hailstone, prior to 1970, fell near Potter, Nebraska, on July 6, 1928 and was 5.5" in diameter.

UPDRAFT SUPPORT OF HAILSTONES

Large thunderstorm updrafts are necessary for the growth of hailstones. If the upper atmospheric temperature is cold, surface heating of the lower atmosphere causes the atmospheric stability to decrease, creating an environment for the development of large thunderstorms with intense updrafts. The updrafts necessary to support hailstones of various sizes have been calculated. Tremendous updrafts are necessary to support the largest hailstones.

Information on the mass per unit volume (density) of hailstones is required in order to calculate the magnitude of the supporting updraft. The density of normal ice is 0.9 gm/cm^3, but hailstones have varying densities depending on the number of milky layers and number of bubbles that are incorporated into them. The average density of hailstones is about 0.8 gm/cm^3. Using this density, an updraft of 60 mph would be required to support a hailstone with a 1" diameter. A hailstone of 5" in diameter requires an extreme updraft of 234 mph to support it (Figure 10-7).

A 5" hailstone would not have to be completely supported by an updraft since the last ring could grow as it fell from the cold air above the freezing level to the surface. However, the hailstone would have to be supported by an updraft of sufficient magnitude to carry it upward when its size corresponded to the next to last layer within the hailstone. Even the next to last layer of larger hailstones includes a very large mass of ice that would require extremely large updrafts within the thunderstorm of more than 100 mph.

ATMOSPHERIC CONDITIONS DURING HAILSTORMS

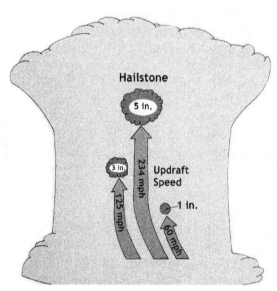

Figure 10-7. Strong updrafts are required in thunderstorms to support large hailstones.

The atmospheric conditions necessary to generate hailstorms are the subject of scientific investigation but many of the factors that cause hailstorm formation are known. One of the important contributing factors is the presence of an incoming cold front. Major hailstorms usually occur in the warm air preceding the movement of a mass of colder air into a locality. Surface streamline confluence near a cold front contributes to atmospheric instability.

The jetstream and upper atmospheric streamline diffluence are associated with hailstone formation. The jetstream is influential in providing flow of air around a large thunderstorm feeding energy into the storm that becomes organized into high speed updrafts and downdrafts. Upper atmospheric streamline diffluence is important in contributing to the updrafts within a thunderstorm. Therefore, both surface and upper atmospheric features are related to hailstorm formation (Figure 10-8).

Considerable wind shear is normally present in the atmospheric environment where hailstorms form. One study showed that the wind speed from the surface to 500 mb was very different from that compared to the wind speed from 500 mb to 250 mb. The difference between the two was related to the size of the hailstones produced. A wind shear of 40 mph between the two layers corresponded to heavy hail, 38 mph to moderate hail, and 30 mph to light hail. A wind shear of 27 mph or less was related to no hail formation. This showed a direct relationship, with greater wind shear more likely to produce large hail.

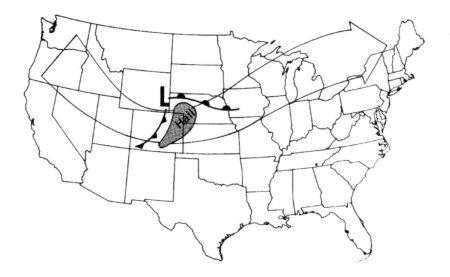

Figure 10-8. Hailstorms are most common in the warm air sector of a frontal cyclone. They are also associated with strong winds in the upper atmosphere.

An unstable atmosphere is associated with hailstorms. A comparison of the atmospheric temperature profile near the time of different hailstorms, as well as the temperature profile near the time of tornadoes, has shown that hailstorms are associated with an unstable atmosphere, although an upper air inversion is not usually present when hailstorms develop. This indicates one of the atmospheric conditions that causes hailstorms to have a different geographical distribution from that of tornadoes.

HAIL PRODUCING THUNDERSTORMS

Hailstorms may be associated with isolated supercell storms, multicell thunderstorms, or squall lines. There is some experimental evidence that indicates that the most severe cell in a squall line or in a multicell group may develop the same organized structure as the single supercell thunderstorm. Most thunderstorms that produce tornadoes also develop hail, although many hail producers do not generate a tornado. The appearance of a thunderstorm that has hail is different from an ordinary thunderstorm. The optical characteristics of sunlight striking a hailstorm cause a dark blue-green color. The very large water droplets and hail in the thunderstorm

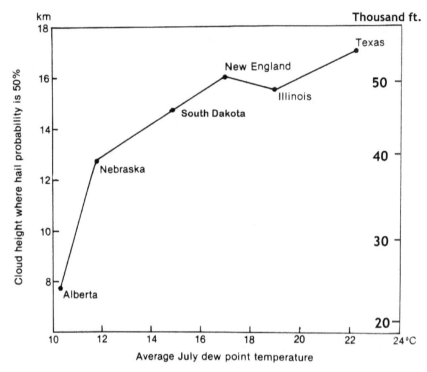

Figure 10-9. Thunderstorms that produce hail are typically about 25,000 ft. in height in Alberta compared to 55,000 ft. in Texas. (After Grosh and Morgan, Preprint Vol. Severe Local Storms, AMS, 1975)

reflect the light in a different way causing the greenish appearance.

An analysis of hailstorms in South Dakota showed that the average top of the thunderstorms was 46,000 ft. based on 134 hailstorms. If a thunderstorm grew to a height of 45,000 ft., the probability was 50% that it would produce hail. If the thunderstorm grew to 55,000 ft. in the late afternoon, it was almost certain to have hail associated with it. The height of thunderstorms that produce hail varies with geographic location (Figure 10-9). A thunderstorm that reaches only 25,000 ft. in Alberta, Canada, has a 50% chance of producing hail while a height of 55,000 ft. in Texas results in a similar probability. Hailstorms originating in locations with warmer and more moist sub-cloud air must reach greater heights to produce hail at the ground.

A severe thunderstorm produces rain, hail, and tornadoes in specific areas within the thunderstorm. Hail normally falls in the central part of a thunderstorm with the major rain area in the leading part of the thunderstorm, typically northeastern, and tornadoes in the

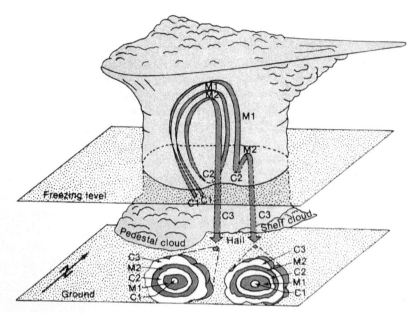

Figure 10-10. The various rings of a hailstone are formed as it circulates within a thunderstorm. Clear layers are formed where the ice freezes very slowly; milky layers are formed under more rapid freezing conditions. The number of trips through a thunderstorm determines how many rings of ice the hailstone will have.

southern part of the thunderstorm. The major hail area in the central part of the storm is related to the thermal updraft and the flow of air within the thunderstorm, causing a hailstone to take one or more paths above and below the freezing level (Figure 10-10). As in the case of tornadoes, most thunderstorms that produce hail come from the southwest. They come from the southwest for the same reason that other severe thunderstorms move in this direction. The upper atmospheric winds are more likely to come from this direction, thereby, determining the path of the thunderstorm. The speed of travel of hailstorms, like tornado producing thunderstorms, is slower than the average winds.

HAIL FORMATION

The nature of hailstone generation by thunderstorms is still under investigation but it is known that hail is generated by a combination of updrafts and downdrafts in supercells. The internal structure may include

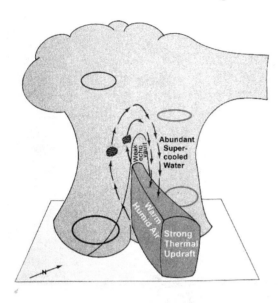

Figure 10-11. Hailstones are produced within a thunderstorm by the appropriate air currents. Within a double vortex thunderstorm, the primary path of hailstones is between the vortices within the strong updraft in the leading front part of a thunderstorm. As a hailstone falls into the rain area, abundant supercooled water is available for making it larger.

a double vortex where the cyclonic circulation is not quite strong enough to generate a tornado. Hail is generated be-tween the two vortices by single or repeated circular trips within the thermal updraft of the thunderstorm that carry the hailstone above and below the freezing level (Figure 10-10). Time spent above the freezing level results in a milky hail layer from rapid freezing, while the hailstone develops a layer of clear ice near or just below the freezing level. Additional layers may be deposited as the thermal updraft carries large quantities of supercooled liquid water above the freezing level to coat the large hailstone. The number of trips above and below the freezing level or number of coatings of supercooled water determines the number of layers within the hailstone giving general information on the path of the hailstone within the thunderstorm. A large hailstone that makes only one trajectory through the thunderstorm can develop successive layers near the freezing level if turbulent motion carried the hailstone repeatedly across the freezing level producing successive layers of milky and clear ice. This possibility combined with coatings of supercooled liquid water delivered above the freezing level by the thermal updraft also explain the layered nature of a hailstone.

Frequently the outer ring of hail is much larger than the other inner layers. This shows that even fairly large hail can stay in the upper part of the thunderstorm for some time acquiring alternating rings of ice. The last clear layer with the greatest mass of ice develops as the

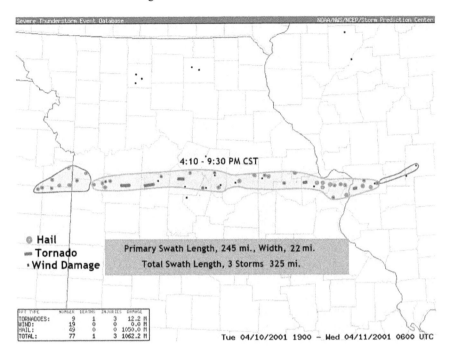

Figure 10-12. The path of the most devastating single hailstorm in the United States. (After NWS analysis)

hailstone falls through the part of the cloud with high water content. A large hailstone may be cold enough to cause freezing of liquid water for some distance below the freezing level.

The circulation of hailstones within a thunderstorm is shown in Figure 10-11. They may be carried upward by the strong thermal updraft to the top of the weak echo vault near its boundary where they are able to take advantage of the abundant supply of supercooled liquid water in the main updraft.

MOST DAMAGING HAILSTORMS

The worst year on record for hail damage in the United States was 2017 with hailstorm damages costing $22 billion. The record for the most damaging single hailstorm was set on April 10 of 2001, when a supercell thunderstorm with hail up to three inches in diameter tracked across Missouri from just east of Kansas City to St. Louis (Figure 10-12). The storm began near Kansas City and produced nine tornadoes, damaging

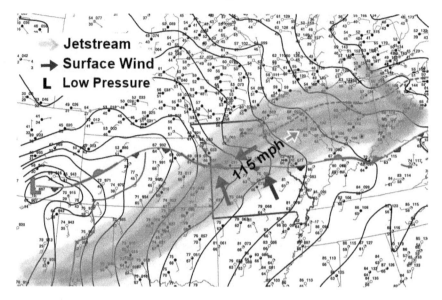

Figure 10-13. The weather patterns for April 10, 2001 included a stationary front extending across Missouri beneath a jetstream from the southwest.

winds and large hail. The weather patterns shown in Figure 10-13 include a stationary front extending across Missouri underneath a strong jetstream from the southwest. The winds at 250 mb were measured at 6:00 pm over south Missouri at 104 mph and 125 mph over north Missouri. We can assume the winds over the thunderstorm in central Missouri were an average of the two at 115 mph from the southwest. The surface winds were converging at the stationary front. This, along with the eastward movement of the thunderstorm at 46 mph, provided the perfect setting for maintaining the internal structure of the severe thunderstorm that was most likely a double vortex thunderstorm. It obviously had a stable internal structure to allow it to survive the strong winds in the upper atmosphere for 5.33 hours.

This storm had a constant and stable cyclonic circulation to allow the Magnus force to pull it to the right of its path of motion at a constant angle to balance the strong jetstream that would have pushed it along a path toward the northeast. The cyclonic circulation was enough to deflect the thunderstorm by almost exactly 35 degrees to allow the storm to move due eastward instead of the direction the jetstream would have pushed it without cyclonic circulation. The two short-path storms before and after

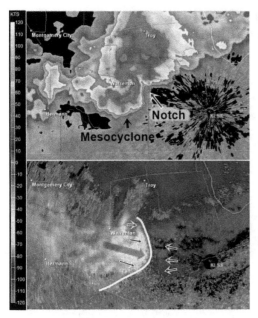

Figure 10-14 Radar reflectivity (upper) and wind velocity for image centered near Warrenton in central Missouri. (NWS photograph)

the long-path supercell shown in Figure 10-12 did not have as much cyclonic rotation since their paths were both more toward the northeast. Support for the double vortex internal structure for this storm is that it survived 115 mph winds from its backside for over five hours. This would only be possible with a double vortex structure as any other configuration could not have survived for five hours in a 115 mph environment. The winds relative to the thunderstorm that was traveling at 46 mph were winds against the lower front part of the storm from the east at 46 mph due to its motion plus the measured additional converging easterly winds. The surface wind at 6:00 pm measured at Columbia Missouri was 23 mph from due east. So, the relative winds at the lower part of the storm were from the east at 46 plus 23 equals 69 mph. The relative winds at the back of the storm were 115 minus 46 or 69 mph. In making this calculation I was amazed that the numbers matched exactly to give a stable environment for this extremely unique storm with matching winds from its back at higher altitudes against incoming winds from the opposite direction at lower levels of the storm. This was the perfect setup for supporting the double vortex structure shown in Figure 6-1.

The radar reflectivity and wind velocities are shown in Figure 10-14. The arc in the southeast part of the storm (top image) most likely corresponds to a well-developed mesocyclone with strong circulation to carry the raindrops around to the east side of the storm where a notch exists. The notch in a double vortex thunderstorm separates the

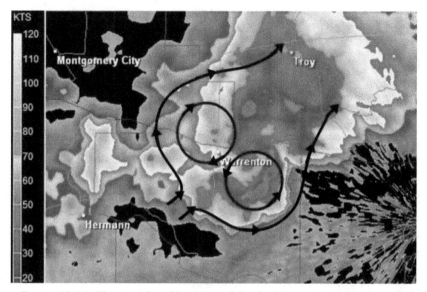

Figure 10-15. The speculated location of the double vortex within the radar reflectivity of the major hail producing storm with its surrounding air flow in upper levels of the storm.

mesocyclone with cyclonic circulation from the rain area to the northeast thereby creating a notch.

It is interesting to speculate on the size and orientation of the double vortex in this storm. This is shown in Figure 10-15 for the upper part of the storm. From Duel-Doppler measurements from a previous storm shown in Figure 5-4 the size of the two vortices are similar and are centered on either side of the main rain area in the thunderstorm. The anticyclonic vortex extends outside the high reflectivity area. With this configuration shown in Figure 10-15 it is easy to see how this storm lasted for such a long time surrounded by winds of 115 mph. The winds around the sides of the storm continued to feed the circulation within the storm and this continued to suck in the warm air in front of the storm as it moved along at 46 mph. The warm air coming into the front of the storm at 69 mph rose upward as shown in Figure 5-6 to block the winds in back of the storm with the same speed. The greater velocity of the cyclonic vortex continued to pull the storm to the right as the heavy rain and hailstones dampened the anticyclonic rotation. Everything about the storm was steady state for five hours as it traveled due east the entire time because of the strong cyclonic rotation even when continually bombarded by 115 mph winds from the southwest.

Total insured losses alone for this hailstorm were estimated to be $2.2 billion. According to an account by the National Weather Service, during the afternoon and evening hours from 4:10 to 9:30 pm the long-lived supercell thunderstorm traveled across Missouri producing catastrophic hail damage. It generated a hail swath with very large hail for a distance of 245 miles with a path width of 22 miles. The supercell moved due east through the highly populated Interstate 70 corridor from Kansas City through St. Louis. It remained centered over Interstate 70 for the entire distance. Most of the hail ranged in size from 1.00-3.00 inches in diameter, however south of the largest hail, marginally severe hail (0.75-1.00 inch) also caused considerable damage as it was propelled by 70+ mph downbursts from the rear flank downdraft winds. This storm has been called the "Tristate Hailstorm" although according to a NWS analysis a second related cell in Kansas must be included for it to cover three states as shown in Figure 10-12. The single supercell that crossed Missouri was the costliest hailstorm in U.S. History.

Insurance claims included 120,000 home claims, 65,000 auto claims, and 8,000 commercial claims. It is believed nearly every home and business in northern St. Louis County suffered hail damage. All of the hundreds of SUVs parked outside at the Ford Motor Company assembly plant in Hazelwood were damaged, while in the adjacent community of Florissant, every home was estimated to have received damage. Twenty-four commercial and military aircraft at Lambert St. Louis International Airport were also damaged. The supercell produced a total of 9 weak tornadoes (6 - F1, 3 - F0) with path lengths ranging from 1 to 10 miles. The F1 tornado which struck Fulton destroyed a mobile home producing the first tornado fatality in Missouri since 1994. With $12 million damage reported from the tornadoes, the total damage from the tornadoes paled in comparison to the hail damage.

SUMMARY

Hailstorms cause considerable damage to agricultural crops each year in addition to damage to personal property and animal life. Hailstorms are the greatest problem in the central Great Plains reaching a maximum in Northeastern Colorado and Southeastern

Wyoming. Hailstorms are most frequent in May and June in the late afternoon hours.

Hailstones are composed of alternating layers of milky opaque ice and layers of clear ice. The thunderstorm environment where the hailstones originated determines the appearance of the ice layer. Hailstones come in various shapes. High speed updrafts are required in thunderstorms to form large hailstones. An updraft of 60 mph is required to support a hailstone of only one inch in diameter. Some important atmospheric conditions that contribute to the development of hailstones are surface streamline confluence ahead of a cold front, high wind speeds within the jet stream, large wind shear and an unstable atmosphere. Hailstorms may occur from any of the three different types of thunderstorms- supercell, multicell, or squall line. Multicell and squall line thunderstorms account for a greater percentage of the total amount of hail, but the hail from a supercell thunderstorm may be larger and more damaging. Hail is most likely to fall beneath the central part of a thunderstorm while heavy rain generally occurs beneath the leading edge of the thunderstorm, and tornadoes, if present, are more likely on the southern edge of a thunderstorm moving toward the northeast. Most hailstorms come from the southwest and travel at a speed slightly slower than the average wind throughout the atmosphere.

The worst year for hail damage was 2017 when hailstorms caused $22 billion damages. The billion-dollar single hailstorm occurred on April 10, 2001 as it traveled across central Missouri from Kansas City to St. Louis. It formed along a stationary front beneath a strong jetstream from the southwest. This supercell produced hail larger than baseballs and nine tornadoes that killed one person and damaged many houses, airplanes and automobiles.

Watch hail formation in a thunderstorm on YouTube. https://youtu.be/pSd-b42Ss6Q

11

The Mighty Middle-Size Storm

A cooperative weather observer, J.E. Duane, located on one of the Florida Keys experienced the hurricane on September 2, 1935, that contained extremely strong winds and the lowest pressure ever measured in the United States at that time. He gave this account:

The barometer reading was decreasing on September 2, as large sea swells and heavy rain continued throughout the afternoon. The temperature was 79°F when the winds increased to hurricane speeds at 6:45 pm. As we crouched in our cottage, a huge 6 by 8 in. beam 20 ft. long came flying through and barely missed three persons. At about 9 pm the winds abated and we could hear other noises besides the wind and rain. The barometer stood at a low reading of 27.2 in. and the light winds told us that we were in the eye of the hurricane. We headed for the last and only cottage that I thought could stand the blow due to arrive shortly. All 20 of us now waited patiently for the hurricane winds to return.

During the lull of 55 min, the skies cleared, with a very light breeze throughout the lull. As I took my flashlight to investigate the sea level only about 50 ft. from the cottage, I saw the water begin to rise very rapidly. As I raced back to regain entrance to the cottage, the water caught me waist deep. I managed to get back into the house and close the door. But soon the house began to sway back and forth and we knew it was floating. At 10:15 pm a blast of wind came at full force from the south-southwest. The house began to break apart. I glanced at the barometer which read 26.9 in. but dropped it into the water as I was blown out to sea. I became entangled in the fronds of a coconut tree and hung on for dear life. I could see the house floating out into the ocean as those still inside flashed their flashlights. I was then struck by some object and knocked unconscious.

I became conscious in the tree a few hours later at 2:25 am on September 3. I was 20 ft. above the ground and all the water had disappeared from the island. I climbed down from the tree and saw that the house had been blown back onto the island. I was amazed to find all the people safe inside. A barometer in the house was showing signs of rising pressure, but very slowly. Hurricane winds continued until 5 am with terrific lightning flashes. Although the whole camp was demolished not a single life was lost.

HURRICANES, TYPHOONS AND CYCLONES

Hurricanes are strong atmospheric vortices that are intermediate in size between the larger frontal cyclones and much smaller tornadoes. These tropical storms are called cyclones in the southern hemisphere and in the Indian Ocean. They are called typhoons in the Pacific and hurricanes in the Atlantic ocean. They originate only in the tropical Tradewinds where the ocean temperatures are quite warm, greater than 80°F. Regions around the world were these storms form are shown in Figure 11-1. The tropical Tradewinds are affected by large high-pressure areas, subtropical highs, that are present most of the time over the oceans at about 30° latitude. In the northern hemisphere these are called the

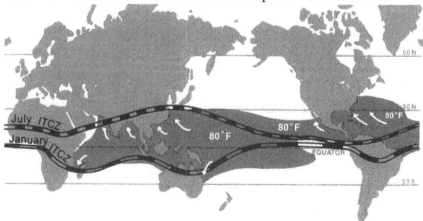

Figure 11-1. The hurricane formation region is governed by the location of warm ocean temperatures greater than 80°F. In addition, hurricanes do not form within about 5° of the Equator because of lack of Coriolis acceleration.

Hawaiian and Bermuda highs. The winds blowing around the high pressure travel anticyclonically in the northern hemisphere and create persistent northeast winds on the southern side of the high-pressure areas that are known as the Tradewinds. It is within these Tradewinds where a cloud mass may be transformed into one of the most devastating storms on earth, not as strong as a tornado but much larger. Hurricane watch areas cannot be forecast on the basis of appropriate atmospheric conditions, as tornado watches can, since the precise atmospheric conditions that cause the development of hurricanes are not known. Unlike tornadoes, however, hurricanes are long-lived and can be spotted on satellite photographs (Figure 11-2) and tracked carefully prior to their arrival onshore.

STAGES OF HURRICANES

The beginning of a hurricane is a small **tropical disturbance** in the Tradewinds. Such tropical disturbances are often present because the high moisture content of the air and the abundance of heating from the ocean surface cause the air to become moist and unstable enough to rise and form clouds. The normal straight flow in the Tradewinds may become curved and form so-called easterly waves with areas of cloud development. As these develop into vortices, the formation of a hurricane is in progress. It is not known why some tropical disturbances suddenly grow into a tropical storm and others do not,

Figure 11-2. NASA'S Aqua satellite captured this unique picture of four hurricanes and smoke from the record setting western fires on Sept. 15, 2020. Hurricane Sandy is making landfall from the Gulf of Mexico.

even though preliminary conditions appear similar.

As the developing hurricane passes to the next stage, called a **tropical depression**, the central pressure begins to fall and the winds begin to flow circularly around the lower pressure (Figure 11-3). One or more closed isobars are typical of this stage with winds less than 39 mph. In the third stage the developing hurricane becomes a **tropical storm** with more closed isobars and lower central pressure with winds gaining speeds of 39 to 73 mph. The **hurricane** stage is reached as the winds become 74 mph or greater. An eye begins to appear characterized by little or no clouds, warmer temperatures, and lighter winds. The central pressure of the hurricane is typically less than 950 mb as the storm matures to its full strength.

ORIGIN OF HURRICANES

Hurricanes are generated as the atmosphere absorbs large quantities of moisture from the ocean. An upper air inversion is usually present over large areas of the Tradewinds as hurricanes develop. The inversion is a region in the atmosphere where the temperature increases with height instead of the usual decrease with height. It is caused by sinking air

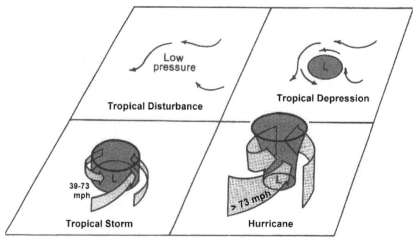

Figure 11-3. As the pressure begins to fall over tropical waters, the air begins to circulate around this area of lower pressure. As the wind speeds increase, the centrifugal force directed outward from the low-pressure center increases, making it more difficult for air to move into the low-pressure center. As this pressure-wind speed relationship intensifies a hurricane is formed.

around the large subtropical high. This inversion contributes to the buildup of a large contrast between the warm, moist air below it and the cooler air aloft.

Atmospheric vortices frequently develop near the intertropical convergence zone, a region where the Tradewinds of the northern and southern hemispheres meet. This convergence zone moves with the season and was shown in January and July in Figure 11-1. It is easier for low pressure to develop as the air rises above a convergence zone. As the winds become circular around an area of low pressure, the inflow of air toward the low-pressure center is prevented. The spiraling winds then form a vertical cylinder extending upward through the atmosphere. Thus, the pressure inside the cylinder at the surface is more like that above than that at the surface on the outside of the cylinder.

An atmospheric vortex can be helped by an outflow of air and associated low pressure at the top of the vortex, as in tornadoes and frontal cyclones, or by the spiraling winds themselves as in hurricanes. As the cloud bands around the eye of the hurricane produce large quantities of condensed water, the latent heat of condensation 2400 J/gm (575 Cal/gm) keeps the rising air currents warm enough to rise even faster and add to this atmospheric heat engine process (Figure

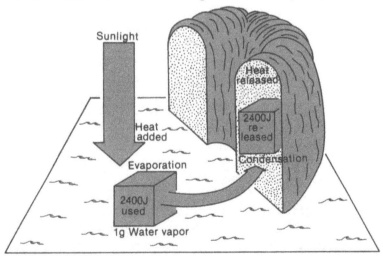

Figure 11-4. A major driving force of hurricanes is the release of latent heat as water vapor condenses within the cloud bands. This sets up a natural heat engine as sunlight evaporates water with the water vapor condensing within the wall of the hurricane. This causes the air to rise faster and is a major driving force for the hurricane.

11-4). The rest of the heat engine is supplied as the sun adds heat to the ocean water and evaporates some of it. The resulting warm moist air feeds the hurricane as it is pulled toward the low surface pressure.

The importance of this part of the heat engine is emphasized by the fact that hurricanes originate only over warm oceans and decay rapidly as they move over land. Hurricanes do not form at the equator even though ocean temperatures are high and the atmosphere is very humid. The region of hurricane formation is from 5 to 30° north and south latitude as was shown in Figure 11-1. This region is not only characterized by warm ocean temperatures, but also has sufficient acceleration coming from the earth's rotation to cause the storm to rotate. At the Equator, the only force due to the earth's rotation is a centrifugal force, whereas away from the equator toward the poles there is considerable spin about a vertical axis because of the earth's rotation. This spin is necessary to start the circular winds of the hurricane.

The formation region of most hurricanes that affect the United States is the Gulf of Mexico or the Atlantic Ocean. Occasionally, hurricanes form in the Pacific Ocean and hit the West Coast. The paths of typhoons, cyclones, and hurricanes are similar in the northern hemisphere as was indicated in Figure 11-1. China and Japan are affected by many typhoons. The Indian coastal regions around the Bay of Bengal have been greatly affected by cyclones because of the density of population on the coast.

In the Southern Hemisphere, tropical storms travel in an opposite direction to those in the Northern Hemisphere. The direction of the surface Tradewinds combined with the winds above influence the direction of a hurricane. As they migrate northward in the northern hemisphere, they are influenced by the westerlies that cause them to curve and travel toward the northeast and then to the east.

SOME FEATURES OF HURRICANES

Hurricanes are generated primarily during the summer and fall seasons of the year in the Northern Hemisphere with peak activity during the month of September. Warm ocean temperatures are so important in their development that temperatures greater than 80°F are required for their formation. This region was shown in Figure 11-1.

Hurricanes do not have weather fronts associated with them as do midlatitude cyclones, but they are circular storms with low central pressure. The structure of a hurricane is quite different from a tornado or midlatitude cyclone; although in the developing stage, the tropical disturbance consists of an area of clouds with ascending air over the low-pressure center. The hurricane is different because it is a surface generated storm much like the dust devil. If you have traveled through the Desert Southwest in the summertime, you have, no doubt, seen dust devils in progress. Some of these are quite small and extend only a few hundred feet upward. They are surface generated as the ground is heated by the sun, and are characterized by rotating air with a downdraft in the center.

Hurricanes are also surface generated, but become a much larger storm, extending through most of the atmosphere to heights greater than 40,000 ft. The eye of the hurricane is a region of light winds, few clouds, and low pressure. Figure 11-5 shows the eye of Hurricane Laura on September 26, 2020, as this hurricane was located just off the coast of Louisiana. The average width of the eye of a hurricane is about 25 mi.

Figure 11-5 The eye of Hurricane Laura is apparent as she approached Louisiana on August 26, 2020 as a category 4 Hurricane. (NASA Photograph)

You might assume that because of the lower pressure the air would rise through the core of the hurricane. However, descending air occurs in the eye of the hurricane because the storm is surface generated with the centrifugal force throwing air outward from the core as it rises. The eye then contains only air coming downward to replace the air that is thrown outward by the vortex. The descending air is warmer than surrounding air because of compression heating and produces cloud-free skies within the eye of the storm.

HURRICANE NAMES

The practice of naming hurricanes after women was begun with the 1953 hurricane season. The name list for hurricanes in the Atlantic and Gulf Coast now has an international flavor because hurricanes are tracked by the public and the weather services of countries other than the United States. Names are selected from library sources and agreed upon during international meetings of the World Meteorological Organization by those nations involved. Male as well as female names were used for the first time in 1978 for Pacific hurricanes, and in 1979 for Atlantic hurricanes.

The National Hurricane Center in Miami, keeps constant watch on oceanic storm-breeding areas for tropical disturbances which may herald the formation of a hurricane. If a disturbance intensifies into a tropical storm, the Center gives it a name from the current list. The six lists of names in the following table are used in rotation and re-cycled every six years, i.e., the 2021 list will be used again in 2027. The only time there is a change in the list is if a storm is so deadly or costly that the future use of its name on a different storm would be inappropriate for reasons of sensitivity. If that occurs, then at an annual meeting by the WMO committee (called primarily to discuss many other issues) the offending name is stricken from the list and another name is selected to replace it. Several names have been retired since the lists were created.

If a storm forms in the off-season, it will take the next name in the list based on the current calendar date. For example, if a tropical cyclone formed on December 28th, it would take the name from the previous season's list of names. If a storm formed in February, it would be named from the subsequent season's list of names.

Hurricane Names for the Atlantic

2021/ 2027	2022/ 2028	2023/ 2029	2024/ 2030	2025/ 2031	2026/ 2032
Ana	Alex	Arlene	Alberto	Andrea	Arthur
Bill	Bonnie	Bret	Beryl	Barry	Bertha
Claudette	Colin	Cindy	Chris	Chantal	Cristobal
Danny	Danielle	Don	Debby	Dorian	Dolly
Elsa	Earl	Emily	Ernesto	Erin	Edouard
Fred	Fiona	Franklin	Francine	Fernand	Fay
Grace	Gaston	Gert	Gordon	Gabrielle	Gonzalo
Henri	Hermine	Harold	Helene	Humberto	Hanna
Ida	Ian	Idalia	Isaac	Imelda	Isaias
Julian	Julia	Jose	Joyce	Jerry	Josephine
Kate	Karl	Katia	Kirk	Karen	Kyle
Larry	Lisa	Lee	Leslie	Lorenzo	Laura
Mindy	Martin	Margot	Milton	Melissa	Marco
Nicholas	Nicole	Nigel	Nadine	Nestor	Nana
Odette	Owen	Ophelia	Oscar	Olga	Omar
Peter	Paula	Philippe	Patty	Pablo	Paulette
Rose	Richard	Rina	Rafael	Rebekah	Rene
Sam	Shary	Sean	Sara	Sebastien	Sally
Teresa	Tobias	Tammy	Tony	Tanya	Teddy
Victor	Virginie	Vince	Valerie	Van	Vicky
Wanda	Walter	Whitney	William	Wendy	Wilfred

PRESSURE

The intensity of a hurricane is related to its central pressure. The lowest non-tornadic atmospheric pressure ever measured was 872 mb (25.75 in.), set on 12 October 1979, during Typhoon Tip in the western Pacific Ocean. The measurement was based on an instrumental observation made from a reconnaissance aircraft. Only a few hurricanes have reached the United States with a pressure less than 900 mb. The first was a hurricane that struck the Florida Keys in 1935 with a pressure of 892 mb (26.3 in). Hurricane Allen in 1980 barely made the list with a pressure of 899 mb. Then Hurricane Gilbert struck in 1988 with a pressure of 888 mb followed by Wilma in 2005 with a record low pressure of 882 mb. (26.05 in.) This pressure is much lower than the traveling

Figure 11-6. The strong winds associated with a hurricane are a very damaging component as shown here by the results of Hurricane Andrew. Many of the houses were completely destroyed by the strong winds. (NOAA photograph.)

frontal cyclones that come through midlatitude locations with accompanying winds and rain. Their central pressure seldom falls below 960 mb. The record low pressure for a frontal cyclone is the storm of January 10, 1993 as it deepened to a central pressure of 912 mb (26.93") between Iceland and Scotland. The extremely low pressure in hurricanes results in strong winds as they try to flow from higher to lower pressure.

WIND SPEED

The wind speed in a typical hurricane is about 100 mph while the greatest that has been measured is twice that speed. Shortly after midnight on October 23, 2015, a group of courageous men and women (Hurricane Hunters) flew into the center of Hurricane Patricia and landed in the history books. With measured winds of 200 mph, Hurricane Patricia became the tropical cyclone with the strongest winds ever recorded anywhere on Earth, with the exception of 231 mph recorded on top of Mt. Washington.

Another storm, Hurricane Allen, has the Atlantic's most powerful recorded sustained winds. The storm, which formed in 1980, had peak sustained winds of 190 mph and a barometric pressure reading of 899 mb.

Understanding Severe and Unusual Weather

Hurricanes Gilbert and Wilma had winds of 185 mph. The amount and type of wind damage depends on the speed. Wind speeds of 50 mph may damage TV antennas or awnings on windows. Higher speeds of 74 mph may cause roof damage to houses. Wind speeds greater than 74 mph begin to produce structural damage to the roofs, eaves or other parts of houses. With wind speeds of 100 mph, major structural damage to buildings is common as was seen previously with the testing of model houses in the wind tunnel. As the wind speeds increase above 150 mph buildings and cities may be reduced to rubble (Figure 11-6).

Measurements show that the highest winds are just outside the eye of the hurricane in a fairly narrow band (Figure 11-7). The speed decreases quite rapidly outward from this band of strongest winds. The wind velocity is very low in the eye, very strong just outside the eye and weakens again farther away from the eye.

SIZE

The size of a hurricane can be measured in two different ways. One method is to use the area within the last closed isobar surrounding the low-pressure center as the outside boundary of the hurricane. On that basis, a composite of 100 storms gave an average size of 300 mi.

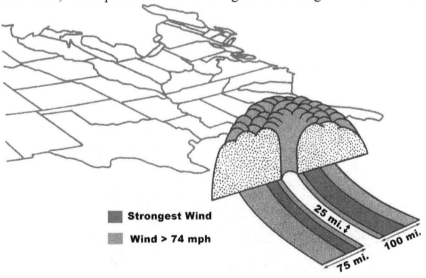

Figure 11-7. The location of the strongest winds and typical size of a hurricane are shown here. The band of strongest winds is just outside the eye of the hurricane. Winds are stronger on the right-hand side of the eye because of the addition of speed of travel of the hurricane.

in diameter for the hurricane. If you consider only the winds that are greater than hurricane velocity, 74 mph, the average hurricane is about 160 mi. in diameter. This combination of size and intensity makes the hurricane a very destructive storm.

SPEED OF TRAVEL AND LIFETIME

The average speed of travel of a hurricane is 15 mph. However, some hurricanes stall, often causing devastatingly heavy rain if this happens when they first make landfall. Other hurricanes can accelerate to more than 60 mph. The average lifetime of a hurricane is nine days. Thus, the typical hurricane moves rather slowly, but lasts long enough to cover a considerable area.

RAINFALL

In addition to the intense winds generated by a hurricane, the amount of rainfall is a major problem. Many million dollars' worth of damage can be done by rainfall amounts of 10 in. or less. An average

Figure 11-8 The storm surge, rainfall and strong winds of a hurricane combine to cause major damages and safety threats. Shown here is mist created by giant waves from an approaching hurricane. (NOAA Photograph)

hurricane produces a tremendous amount of water. Most cities do not have drainage facilities that can handle much more than 4 in. of rainfall per hour.

The rain in hurricanes is most intense in the spiraling bands around the eye. Thunderstorms are embedded in these spiraling bands and produce heavy rainfall. They also have lightning and thunder associated with them.

As a hurricane approaches land it brings warm, humid air that is usually sufficiently different from the air located over the land to create a weather front to add to the deluge of water (Figure 11-8).

OCEAN SWELLS AND STORM SURGE

Still another extremely damaging feature of a hurricane is the ocean swell or **storm surge**, an increase in ocean level that it produces at considerable distance from the storm. The high wind speeds within a hurricane can generate waves to heights of 10% of the speed of the winds in kilometers per hour for wave heights expressed in meters. Such waves produce ocean swells characterized by undulating action of the ocean in advance of the storm center. Ocean swells are produced by the hurricane winds and are propagated outward from the area where they are produced. The winds on the right-hand side of the hurricane generate swells that move in the same direction as the hurricane. The strongest winds are on the right-hand side and produce the greatest swells which may reach the shore a day or more in advance of the hurricane itself.

When the hurricane arrives it also brings a higher ocean level because of the inverted barometer principle. A hurricane with a pressure of 900 mb will have an increase in ocean level of more than 3 ft. because of the lower atmospheric pressure.

Many of the rises in sea level for hurricanes in the Gulf of Mexico have been more than 10 ft. above normal sea level. One of the greatest recorded storm surges was generated by Hurricane Katrina on August 29, 2005, which produced a maximum storm surge of more than 28 ft. in southern Mississippi, with a storm surge height of 27.8 feet in Pass Christian. If you have driven along the coastal areas of the Gulf, you have seen that a rise of this magnitude would inundate a large region.

In the Bay of Bengal off the coast of India, in Bangladesh, even higher waves are generated. The configuration of the bay allows water to be driven into the narrowing portions to cause large increases in ocean levels. Tropical cyclone Bkola in 1970 brought an increase in ocean level within the Bay of Bengal of 34 ft. Because of the dense population of these counties, such tropical storms take a heavy toll of human life, 500,000 for Bhola.

TORNADOES IN HURRICANES

Hurricanes add to their damage potential by producing tornadoes with even higher wind speeds. They are contained within the large thunderstorms that form the spiraling bands in advance of the eye of the hurricane. There is no preferred time of day for the formation of tornadoes within hurricanes. The rotation of the hurricane provides enough energy for individual thunderstorms to generate tornadoes without contributions from surface heating and many of the other factors required outside the hurricane to form severe thunderstorms.

Tornadoes with hurricanes are generally less severe than other tornadoes. Not only are tornadoes that are generated by hurricanes less intense, but they have shorter and narrower paths as well as shorter lifetimes. Other differences include the absence of the temperature inversion that is usually a factor in the generation of tornadoes in the Great Plains and no sharp vertical moisture stratification, but a lot of wind shear.

A comparison of many tornadoes produced within hurricanes shows that the forward right-hand quarter of the hurricane is the most likely location for the generation of tornadoes (Figure 11-9).

Numerous cases of tornadoes associated with hurricanes have been documented, although many more probably occur than can be recorded. Hurricane Beulah produced a record number of 141 confirmed tornadoes mainly in Texas in 1967. Only two other storms have generated more than 100 tornadoes and these were Ivan and Frances in 2004 with 117 and 101 tornadoes.

DISSIPATION OVER LAND

Hurricanes dissipate very rapidly after landfall. The less humid air and rougher terrain drastically cut down the conversion of water vapor to cloud droplets with a reduction in the amount of latent heat to drive the hurricane. Thus, when they strike land, they lose power very rapidly. The intensity of the hurricane drops by 50% by the time it is

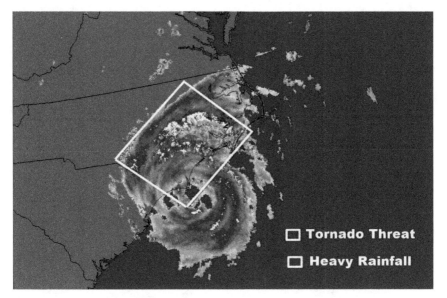

Figure 11-9. The threat of tornadoes is greatest in the front right-hand quadrant of a hurricane. (NOAA Photograph)

only 150 mi. inland. If the winds in the hurricane were 150 mph when it struck the shore, they would have decreased to 75 mph after moving 150 mi. inland. Coastal areas are therefore affected most by hurricanes and hurricane velocity winds do not reach inland more than about 150 miles.

The damage potential decreases with distance from the shore with 90% of the damage done within 60 mi. of the ocean, 50% a little farther inland, and no damage by the time the storm reaches as far inland as Oklahoma or Arkansas, for example. The damaged area along the coast and its extent inland is shown in Figure 11-10 for a typical hurricane.

The coastal regions that experienced inundation of water from Hurricane Carla are also shown. This factor causes much additional damage in low-lying areas. More damage typically occurs on the right-hand side of the path of the hurricane than on the left. If the hurricane is moving northward at the average speed of 15 mph, then the winds on the right-hand side will be 30 mph faster than those on the left-hand side because the forward speed adds to the winds on the right-hand side and subtracts from those on the left-hand side. This effect is significant and can be seen in damage paths. Captains of ships used this information for decades, before the age of satellites, to avoid the strongest winds in a hurricane.

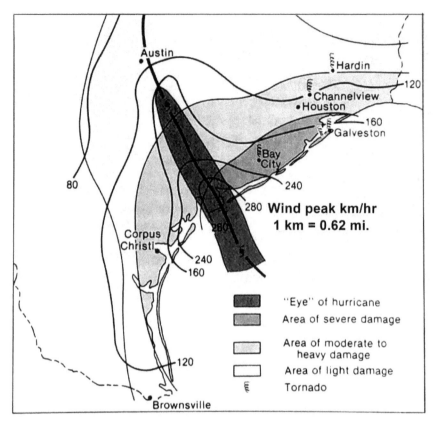

Figure 11-10. The damage areas from Hurricane Carla are typical of those associated with a hurricane. Damages occur from inundations of coastal areas from high water levels and high wind speeds. Damages decrease farther inland and extend farther on the right side.

SUMMARY

Hurricanes are atmospheric vortices intermediate in size between frontal cyclones and tornadoes. Hurricanes originate as tropical disturbances over warm oceans within the Tradewinds. The tropical disturbance intensifies into a tropical depression, then into a tropical storm, and finally becomes a hurricane as the winds increase to 74 mph. Hurricanes originate near the Intertropical Convergence Zone with the latent heat of condensation providing the energy for driving the storms. They form over warm oceans from 5 to 30° N and S latitude where the water temperature is greater than 80 degrees.

The structure of the hurricane includes an eye about 25 mi. in diameter with low pressure and descending air. The lowest recorded pressure in an Atlantic hurricane was in Wilma with a pressure of 877 mb. The typical hurricane has winds of at least 74 mph over an area 160 mi. in diameter. Hurricanes typically travel at 15 mph and last for 9 days. They produce heavy rainfall and large ocean swells, but decrease in intensity quite rapidly after landfall.

Hurricanes may produce tornadoes within the leading rainbands. They are generally smaller than average but add to the damaging effects of these storms. Another damaging characteristic of hurricanes is the large ocean swells that they produce. These have been as great as 28 ft. in the Gulf of Mexico or 34 ft. in the Bay of Bengal.

12

Categories, Prediction and Notable Hurricanes

According to the computer drawn map we would be entering the outer rainbands shortly. Verification came rapidly in the form of extremely turbulent air currents, heavy rain, and electrical displays that surrounded and buffeted our reinforced and specially equipped aircraft. Our carefully prepared statistical procedures dictated that we begin the measurements directed by our research activities in unison with pressure, temperature, and wind measurements. I could hardly believe that we were actually going to fly through the strongest part of this hurricane. As the 100 mph winds engulfed us, we read the first instructions. Even as we pulled the appropriate levers and recorded numbers it was impossible to miss the sudden change in the outside environment as we broke into clear air and calm winds. There was no mistaking the circular wall of clouds that now surrounded us as we flew across the eye of the hurricane and prepared again for the fury awaiting us along any escape route we might choose. But such is the life of a Hurricane Hunter who works for the National Hurricane Center.

HURRICANE CATEGORIES

The naming of hurricanes was covered in the previous chapter. In addition to naming hurricanes the National Hurricane Center assigns an intensity category to each hurricane based on its wind speed.

Saffir- Simpson Hurricane Wind Scale

Category	Sustained Winds	Types of Damage
1	74-95 mph	**Very dangerous:** Damage to roof, siding and gutters. Damage to power lines.
2	96-110 mph	**Extremely dangerous:** Major roof and siding damage. Trees will be snapped. Near-total power loss.
3 (major)	111-129 mph	**Devastating damage:** Removal of roof. Trees will be uprooted. Electricity and water will be unavailable.
4 (major)	130-156 mph	**Catastrophic damage:** Severe damage, loss of some exterior walls. Fallen trees will isolate residential areas.
5 (major)	157 mph or higher	**Catastrophic damage:** Homes will be destroyed. Power outages will last for weeks to possibly months.

Assigning an intensity category to a hurricane is important so that residents of coastal regions will have a better understanding of the severity of an approaching hurricane and can respond appropriately. It also provides a better reference to past hurricanes since surviving through a category 1 hurricane is nothing like what comes with a category 5 hurricane.

PATHS OF HURRICANES

The typical movement of Atlantic hurricanes is initial travel from the southeast. As they move into more northerly latitudes, they generally start to curve with a more northward movement, followed by movement to the northeast. As hurricanes develop in the warm tropical waters of the Atlantic Ocean they generally take a curved path as shown in Figure 12-1. They may travel westward for a week, thus endangering the eastern coast of the United States even if they originate several hundred miles away. If a hurricane develops in the Gulf of Mexico it will almost certainly hit the Gulf Coast.

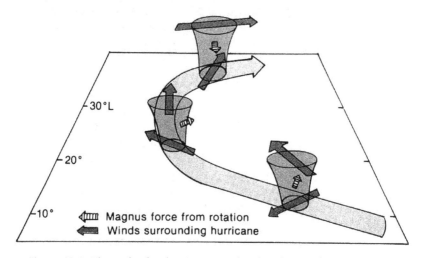

Figure 12-1. The path of a hurricane is related to the winds surrounding the hurricane including those at low and upper levels. The rotation of the hurricane also causes a force to the right of the path of motion. Various combinations of these forces cause different hurricane paths.

Many hurricanes do not have a typical path. A hurricane may suddenly change directions and even follow a looping path. This may cause it to strike land when its previous path would have kept it over the ocean. Hurricane Betsy was of this type. She was generated out in the Atlantic and maintained hurricane winds for over nine days before she struck the Gulf Coast. It first appeared that Betsy would reach the Eastern United States, but a looping path took her around the tip of Florida and into the Gulf instead. Such hurricanes that describe a looping pattern as they move northward make it very difficult to predict their future movement while they travel in a circle.

The distribution of high- and low-pressure centers affects the paths

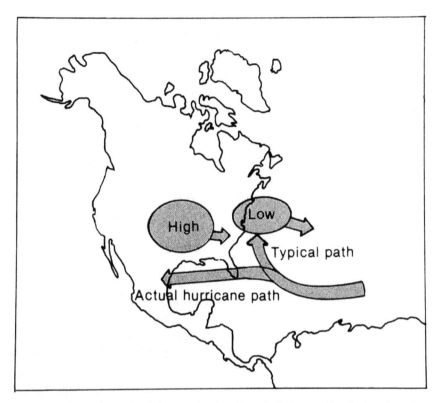

Figure 12-2. The path of a particular hurricane is influenced by the location of travelling high- and low-pressure centers. Hurricanes tend to avoid either high- or low- pressure centers.

of hurricanes as shown in Figure 12-2. Hurricanes will change their course to avoid either a high- or low-pressure system in midlatitudes. You might think that low pressure areas would attract a hurricane but it will go around either a midlatitude cyclone or high-pressure system. Hurricanes will travel either eastward or westward to avoid a midlatitude cyclone or anticyclone located to their north and do not ordinarily combine with either because their structure is so different.

HURRICANE PATH PREDICTION

About 70% of the hurricanes behave in typical fashion by traveling in paths that make them easier to predict. It is the other 30%, which includes hurricanes such as Betsy, that make predictions of future paths of hurricanes difficult. Another difficulty in providing precise warnings to people living in coastal areas is that hurricanes may only develop strong winds within the Gulf of Mexico shortly before landfall.

Most people who have experienced a hurricane have been in its fringe area, since the most intense winds are confined to a fairly small area. Therefore, they may have a false impression of the severity of a major hurricane. This impression is reinforced by weaker hurricanes that strike land. This sets up situations such as that when 25 people held a hurricane party as Camille blasted the Richelieu Apartments in Pass Christian, Mississippi. Only two of them survived; one of them, a boy, floated out the window on a mattress. Both he and the other survivor were severely injured by the high winds, waves, and debris.

The National Hurricane Center in Miami uses many different track prediction models in an effort to pinpoint the location of landfall for hurricanes reaching the United States. These may be divided into four basic types (Figure 12-3). One type is purely dynamic; the others are statistical or climatological. If the location of landfall can be predicted, it makes a tremendous difference in warning people of the approaching danger. The accuracy of predicting the exact location of the center of the hurricane at landfall may only be within 50 mi. depending on its distance from shore. But this distance may mean that one particular city is affected instead of another. Timely warning is very important in evacuating a large city in the path of a major hurricane. None of the techniques predict the hurricane path perfectly every time. This is the

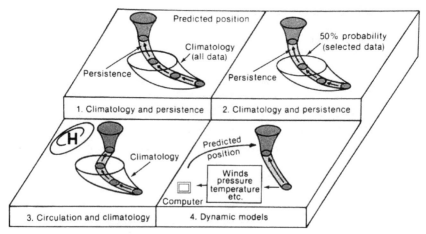

Figure 12-3. Hurricane track prediction techniques include the four different methods illustrated above. These are based on climatology, persistence, circulation, and dynamic models.

Understanding Severe and Unusual Weather

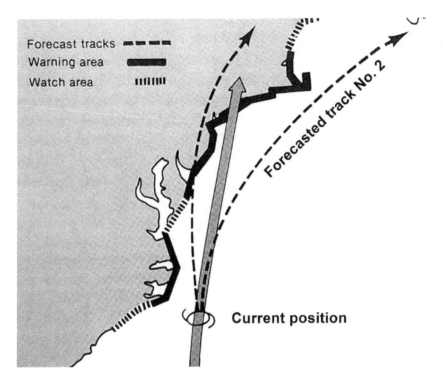

Figure 12-4. Forecasting the track of a particular hurricane is very important since several hundred thousand people must be evacuated from the path of a typical hurricane. The forecasted path determines which regions get watches or warnings.

reason for having different prediction models to pinpoint as specifically as possible where the hurricane is going to reach land.

One of the hurricane path prediction models used by the National Hurricane Center is based on climatology and persistence. Persistence is simply the behavior of the storm in the past. This includes speed of travel and its direction during the past. This direction and speed is then used to project the hurricane speed and direction for the next day. The climatology of previous hurricanes including the speed and direction of all other hurricanes is added to form this technique for predicting the movement of a particular hurricane. This gives a prediction for the location of the storm several hours in advance. Figure 12-4 shows the predicted path and affected coastal area for a particular hurricane. The specific path is so important because a slight variation may affect many people as shown in this figure.

Other methods used at the National Hurricane Center are similar to the previous technique except that only data near the particular location of the current hurricane are used. Then the projection is based on hurricanes located in the same area combined with the persistence of the current hurricane.

Other techniques used are based on circulation and climatology. The climatology of past hurricanes is combined with such factors as the surrounding pressure systems, upper air flow patterns, direction of the trade winds, and other information. Many of the methods consist of dynamic models. The computer processes equations that require information concerning winds at the surface and above. Thus, data on the current atmospheric conditions are needed to compute the location of the center of lowest pressure (hurricane center) at a later time.

No one of the many different hurricane path prediction models is consistently more accurate than the others. All are used for each hurricane that forms and the forecaster at the National Hurricane Center uses experience to make the final path prediction. Later evaluations of the actual path with the forecast path have shown that the published path made by the forecaster beats any one of the computer-generated paths.

GALVESTON HURRICANE OF 1900

The record for greatest number of deaths from a hurricane occurred before the age of hurricane track predictions. It belongs to the Great Galveston Hurricane of 1900. Charts of the Gulf of Mexico made long ago show three great oak trees as the landmarks of Galveston, Texas. For three centuries storms in the area battered those big trees, harming them very little. High winds and tides caused some destruction, but never enough to slow Galveston's development or uproot the sturdy oaks. Then on September 8, 1900, a hurricane swept over the island causing catastrophe and felling the great oaks. The death toll of more than 6000 places this storm as the greatest natural disaster in this nation's history.

It was not until the evening of September 7 that a storm warning flag was put up. This was the only hint, besides a very small article in the "Galveston News," that a storm was coming. The "brick dust" sky which traditionally heralds the approach of a hurricane was also missing that day. Everyone knew the city was almost totally helpless against the sea. There was no high ground, as the highest point of the city was less than 10 ft. above sea level.

Isaac Cline, head of the U.S. Weather Bureau office at Galveston, had started to monitor the storm at 5 am on Saturday, September 8, not knowing that the hurricane would zero in on Galveston. There had been light rain early in the morning then it started to pour. By 3 pm, over half the city was under water with many people trapped in their low-lying houses. The exact speed of the wind during the fifteen hours that it battered Galveston is not known since the Weather Bureau's anemometer blew away after registering 84 mph at 5:15 pm. It has been estimated to have been a category 4 with winds reaching at least 145 mph blowing masses of salt water across the island.

The four bridges to the mainland had washed out early, and all contact with the rest of the world was cut off by midafternoon. The city's water and electric plants ceased to function. By late afternoon there was water over every foot of the island, with the center, the highest part, under more than 3 ft. of debris-filled water. At 8:10 pm, the barometer stood at 28.53 in., the lowest it had ever been in Galveston. Many people sought sanctuary in churches and hospitals that collapsed, killing almost everyone inside. At about 10 pm, the raging winds abated, and for a few moments the survivors had a tremendous sense of relief, but they were to suffer more. As the wind, which had carried huge quantities of water onto the island, died, a massive surge of waves headed seaward, creating more destruction and drowning many people who survived the hurricane's original assault.

Over 1500 acres of the city had been devastated along with well over 3800 houses. There was an estimated $30 million of property damages. The death toll is listed at 6,000, although discrepancies place the estimated total as high 12,000, based on evidence that many neighborhoods were destroyed and left no friends who could furnish names. Added to this fact is that many strangers and dock workers were in the city with no local ties. It was also impossible to get the precise number of people in some families.

The weeks following the hurricane were grim times, but the Galvestonians would not consider outsiders' advice to abandon the island. Following an incredibly successful long-term rebuilding program, Galveston incorporated such measures as raising the entire grade of the city from 3 to 15 ft. above the old level, constructing a seawall along the Gulf (Figure 12-5), and connecting the island and mainland by a fine trafficway. By 1905 the seawall was 3 mi. long

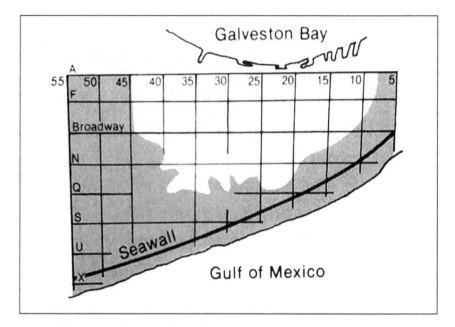

Figure 12-5. Location of the present seawall in Galveston in comparison to the total destruction (shaded area) from the hurricane of 1900. Property south of the seawall (black line) was completely washed away. (After Weatherwise, 32:4, 1979)

and was 18 ft. above the sea level while extending 9 ft. above the streets. The seawall was eventually extended to its present 11 mi. length.

Proof that the wall provided safety was realized as soon as 1915 when a great hurricane as strong as the one in 1900 killed over 250 in two states but killed only eight in Galveston and flooded the city with only 5 ft. of water. The new trafficway went down, all except one section, which held a stalled two-car trolley filled with frightened passengers who survived. The improvements have provided protection several times since, particularly during Hurricane Carla in September 1961. Thus, Galveston remains a beautiful commercial and pleasure port.

HURRICANE BEULAH

Hurricane Beulah of September 1967 has the distinction of producing more tornadoes than any other hurricane. It should also be mentioned that the name Beulah was used previously for a hurricane in August 1963. Although the first Beulah did not strike land, it is also

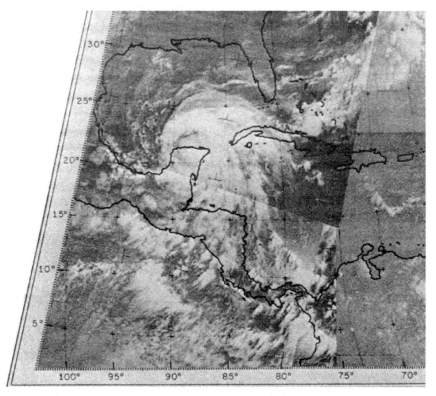

Figure 12-6. Hurricane Beulah September 17, 1967. (NOAA, ESSA Photograph)

noteworthy since it was one of the few hurricanes seeded for purposes of modification.

The weak tropical depression that was to become hurricane Beulah in 1967 formed on September 5 east of the Windward Islands. Hurricane winds were developed by the next day and reached 143 mph prior to touching the southern coast of the Dominican Republic on September 11. Beulah caused considerable damage among the Caribbean Islands before reaching the Gulf of Mexico where she could regain the strength lost because of the rougher terrain of the Islands and loss of warm, humid air from over the ocean. Before landfall on September 20 just east of Brownsville, Texas her pressure was 923 mb with winds of 120 mph. Satellite photography of hurricanes was in its infancy in 1967 as can be seen in Figure 12-6.

Property and crop losses in Texas were about $200 million. Fifty-eight people died in Hurricane Beulah, fifteen in Texas. Five of these were killed by tornadoes. At least 115 tornadoes were spawned by

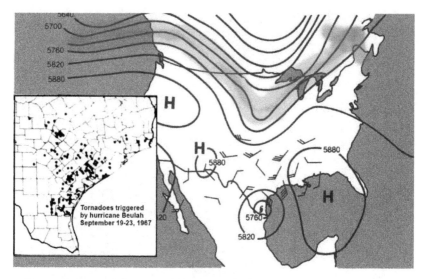

Figure 12-7 As Hurricane Beulah approached Texas on September 19, 1967, the airflow at 500 mb was from the southwest over much of Texas. As the storm reached southern Texas, two days later these winds were replaced by easterly winds north of the hurricane center. This wind shear may have played a role in the formation of many tornadoes.

Beulah (Figure 12-7). Most of the tornadoes were small and formed over rural areas where they caused only minor damage. There were exceptions, such as the tornado that struck Palacios on the morning of the 20th. It picked up ten people and carried them up into the tornado. As they were blown out into a field and dropped, four of them were killed and the others were injured. A tornado also struck Burnet and moved across the city with resulting damages of $100,000. Damages of at least this magnitude were experienced in several other communities.

The path of Hurricane Beulah contributed to the large number of tornadoes that were formed. As Beulah approached Texas and turned toward the west the strong east winds ahead of the eye, carrying tons of moisture, interacted with and then replaced the general westerly winds over Texas (Figure 12-7). Such opposing air currents are sufficient to generate tornadoes as has been demonstrated in the laboratory. Thus, Beulah became the best tornado generator in the history of hurricanes.

Figure 12-8. Hurricane Camille in the Gulf of Mexico August 17, 1969. (NOAA, ESSA Photograph)

HURRICANE CAMILLE

Only a few hurricanes have struck the United States with a category 5 intensity. These are the Labor Day Hurricane of 1935, Camille in 1969, Andrew in 1992 and Michael in 2018. Camille was the first since the age of satellites and better technology and will be described now. Then we will look at the more recent Michael.

Hurricane Camille was the third of its season and has been called the greatest storm to strike the United States. On August 17, 1969, she slammed into the Gulf Coast and reminded man, once again, of his virtual helplessness in a full-scale hurricane.

Camille began as a small rainstorm and grew into a hurricane off the south coast of Cuba. Then, she roared into the Gulf of Mexico gathering fantastic power. The release of latent heat was very effective in driving the heat engine of hurricane Camille as currents of moisture rich air gave her incredible power; and she was fed by these air currents right up to the time she hit land (Figure 12-8).

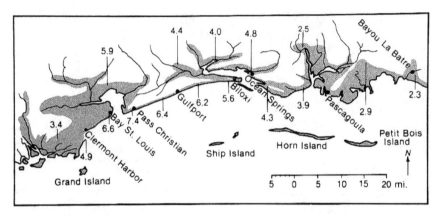

Figure 12-9. Storm surge and flooded areas as a result of Hurricane Camille. Water levels greater than 6 m above sea level were recorded from Bay St. Louis to Gulfport. This was the greatest storm surge ever recorded for the United States. (Climatological Data, National Summary.)

An Air Force reconnaissance crew penetrated Camille's eye on August seventeenth. They radioed back a surface pressure of 901 mb (26.61 in) and 143 mph winds. The report prompted the Director of the National Hurricane Center, Robert Simpson, to state: "Never before has a populated area been threatened by a storm as extremely dangerous as Camille." A central pressure lower than Camille's had been recorded only once before in the North Atlantic-in the Labor Day Hurricane of 1935 near the Florida Keys. In size Hurricane Camille was small compared to other hurricanes. Hurricane winds (greater than 74 mph) extended over an area only 95 mi. in diameter as Camille moved ashore. Only small storms such as Inez of 1966, had hurricane winds extending over smaller areas.

Noted for their devastating storm surges, Gulf Coast hurricanes bring with them large amounts of water. Camille generated ocean levels that were more than 15 ft. above mean sea level from Bay St. Louis to Biloxi. At Pass Christian, Mississippi the water level increased about 7 m (23 ft.). Storm surge levels along the coast are shown in Figure 12-9. From southeastern Louisiana to Biloxi, Miss., almost total destruction was brought to waterfront areas from the combination of severe winds and devastating tides. Hurricane Camille became the most damaging storm in history with property damages of $1.42 billion. Cars and trucks were smashed like toys and giant freighters were tossed about and beached. One of those viewing the devastation described it as having the same appearance as if a giant bulldozer had leveled the whole coastal area.

A first-person account told 50 years after Camille follows:

"I will never, ever, forget the devastation Hurricane Camille did to our pretty little town and all the lives that were lost. In fact, my parents' house was a block from the infamous Richelieu Apartments where the so-called hurricane party took place and so many people lost their lives. My mother had just moved her friends, Zoe and Jack, into those apartments that weekend. Zoe was confined to a wheelchair and both of them perished. Her wheelchair was found but I don't think their bodies ever were".

Hurricane Camille produced tremendous amounts of rainfall. The Gulf Coast area received between 4 and 12 in. Most hurricanes do the majority of their damage along the coast where they hit. As Camille moved inland, she was expected to die a normal death. However, as the center of the remaining low pressure moved into Kentucky and the upper part of the storm interacted with the jetstream it turned abruptly toward the east and began dropping torrential rains. As the modified Camille traveled over the Appalachian Mountains flash floods were common (Figure 12-10). Within a twenty-four-hour period Massies Mill,

Figure 12-10. Hurricane Camille produced disastrous flooding of the James River. The streets of Richmond, Virginia, were navigable only by boat as shown in this photograph. (Courtesy of Richmond Times Dispatch.)

Virginia, received an unbelievable 27 in. of rain. On the 19th of August, Camille brought up to 2 ft. of rain over the James River basin flooding it, and almost doubled the death toll. Camille caused 144 deaths mostly from drowning along the Gulf Coast, while 114 deaths occurred in Virginia and West Virginia from flash floods. An incredibly large number of families (77,985) suffered losses from the hurricane.

HURRICANE MICHAEL

Hurricane Michael of October 2018 was a very powerful and destructive hurricane that became the first Category 5 hurricane to strike the contiguous United States since Andrew in 1992. In addition, it was the third-most intense Atlantic hurricane to make landfall in the contiguous United States in terms of pressure, behind the 1935 Labor Day hurricane and Hurricane Camille in 1969. It was the first Category 5 hurricane on record to impact the Florida Panhandle, the fourth-strongest landfalling hurricane in the contiguous United States, in terms of wind speed, and the most intense hurricane on record to strike the United States in the month of October. The infrared satellite photograph of hurricane Michael is shown in Figure 12-11.

Hurricane Michael made landfall as an unprecedented Category 5 Hurricane in the Florida Panhandle region with maximum sustained wind speeds of 161 mph and a minimum pressure 919 mb. The storm caused catastrophic damage from wind and storm surge, particularly in the Panama City Beach to Cape San Blas areas. The widespread damage spread well inland as Hurricane Michael remained at hurricane strength into southwest Georgia.

At least 74 deaths were attributed to the storm, including 59 in the United States and 15 in Central America. The smaller number of deaths from this category 5 hurricane than Camille is a testament to the effectiveness of public warnings along with the timely response of many thousands of people.

Michael caused an estimated $25.1 billion (2018 USD) in damages, including $100 million in economic losses in Central America, damage to U.S. fighter jets with a replacement cost of approximately $6 billion at Tyndall Air Force Base, and at least $6.23 billion in insurance claims in the U.S. Losses to agriculture alone exceeded $3.87 billion.

Understanding Severe and Unusual Weather

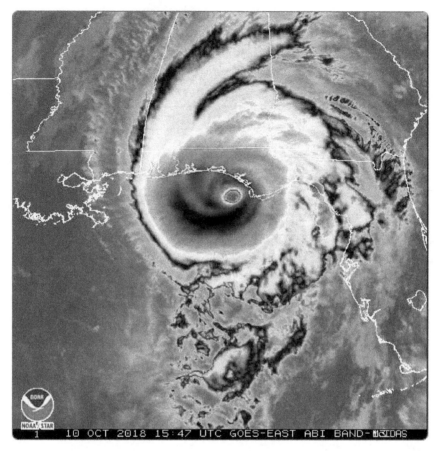

Figure 12-11. Infrared satellite photograph of Michael as it reaches the Florida panhandle as a category 5 hurricane. (NOAA photograph)

HISTORICAL ATTEMPS TO MODIFY HURRICANES

The winds and water accompanying a hurricane are so damaging that considerable interest has existed in modifying these features. Project Stormfury was a national effort to seed hurricanes and measure any resulting modifications. Hurricanes were seeded by adding silver iodide or dry ice to them by means of an airplane. Dry ice crystals have a very low temperature of -78°C and are capable of generating many ice crystals in a supercooled atmosphere. Silver iodide is a substance that has crystals so similar in shape to the ice particles that it also causes the generation of many ice crystals.

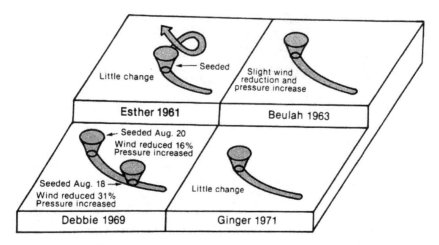

Figure 12-12. Results from seeding four hurricanes. The most positive results were for Hurricane Debbie.

Four hurricanes were seeded by adding seeding agents to the hurricane in particular locations. Two different locations were used. One train of thought was to seed close to the eye of the hurricane to convert more of the water vapor to ice crystals to widen the eye and weaken the storm by conservation of angular momentum as previously discussed. This principle has evident from the fact that the most damaging hurricanes are generally the smaller ones like Camille and Agnes. The attempt at seeding was to cause the diameter to increase, thereby decreasing the power of the winds.

The other effort was to seed several miles outward from the most intense winds to increase the latent heat and rainfall from the warm moist air traveling inward to feed the clouds around the eye wall. If these wind currents were seeded and converted into large updrafts this could cut down on the amount of moisture that would be carried into the most intense region of the hurricane. This would be expected to reduce the most damaging winds near the eye.

Four hurricanes were seeded from 1961 to 1971 (Figure 12-12). The first was Hurricane Esther in 1961. Following this seeding, measurements indicated that more ice crystals were produced. The other observed characteristic was that it began to describe a looping path and struck the United States. There is no direct evidence that the seeding was the cause of the change in direction since none of the others that have been seeded showed a similar change in direction. Since it struck the United States it resulted in new rules for seeding hurricanes. They were required to be much farther away from land when they are seeded.

Hurricane Beulah (not Beulah of 1967) was seeded in 1963 with some indication of weakened winds. After a second seeding some decrease in windspeed was noted. The next hurricane to be seeded was hurricane Debbie in 1969. Seeding on August 18 was followed immediately by a 31% reduction in winds. It was not seeded on 19 August and the winds returned to their original intensity. Another seeding on the 20th resulted in winds dropping by 60%. These measurements gave the best indication that seeding hurricanes might be beneficial.

The fourth hurricane, Ginger, was seeded in 1971. This storm had large oscillations of wind speed and cloud conditions both before and after seeding. Seeding produced no changes greater than the natural variations. Project Stormfury was terminated following these hurricane seeding operations.

Other suggestions for modifying hurricanes have been made such as adding carbon dust to a hurricane or spreading a monomolecular film over the ocean in the vicinity of a hurricane. But these have remained only suggestions and no recent attempts have been made at hurricane modification.

HURRICANE PREPAREDNESS AND SAFETY

If hurricanes are a threat, you should enter the hurricane season prepared. Each spring, recheck your supply of boards, tools, batteries, nonperishable foods, and other equipment you will need if a hurricane strikes your town.

When your area is covered by a **hurricane watch**, continue normal activities, but stay tuned to radio, television or cell phone for National Weather Service advisories. Ignore rumors. A hurricane watch means a hurricane may threaten an area within 24 hours. Continuously monitor the storm's position through Weather Service advisories. Check battery powered equipment. A cell phone (Figure 12-13) or portable radio may become your only link with the outside world. Emergency cooking facilities and flashlights will be essential if utilities are interrupted.

Have your car fully fueled. If you own a boat, secure it before the storm arrives or move it to a safe area. When the boat is moored, leave it. Don't return to it once the wind and waves are up.

When your area receives a **hurricane warning**, additional activities

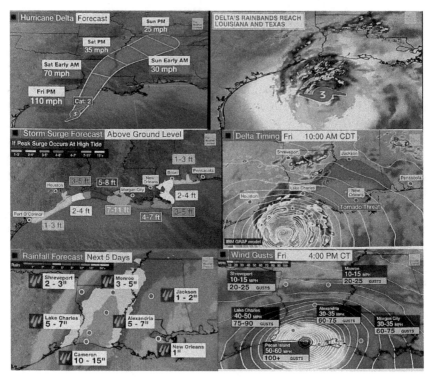

Figure 12-13. An amazing amount of information is available on cell phones from apps such as the Weather Channel app prior to a hurricane landfall. Here are six images obtained on a cell phone prior to landfall of Hurricane Delta on August 19, 2020

should be completed. A hurricane warning means a hurricane is expected to strike an area within 24 hours. Continue to monitor the storms approach. It is time to board up windows or protect them with storm shutters. Secure outdoor objects that might be blown away or damaged, or bring them inside. Store drinking water; your town's water supply may be contaminated or diminished by hurricane floods. Leave low-lying areas when advised to do so. If you live in a mobile home, leave it for more substantial shelter. Mobile homes are extremely vulnerable to high winds. If your home is sturdy and at a safe elevation, remain indoors during the hurricane. Because hurricanes often cause severe flooding as they move inland, stay away from the banks of rivers and streams. Tornadoes are often spawned by hurricanes and are among the storms' lethal effects. Therefore, when a hurricane approaches, watch your cell phone or listen to radio and TV for tornado warnings.

Hurricane watches and warnings are extremely important in reducing the loss of life from these storms. Since several hundred thousand people must be evacuated as major hurricanes approach land, specific evacuation routes are important. Such predetermined travel routes have been planned for many coastal communities with the help of the National Weather Service. Advanced preparation for hurricanes should also include a knowledge of the recommended routes to safety.

SUMMARY

Each hurricane in the Atlantic since 1953 has been given a name. Hurricanes are assigned a category based on wind speed and expected damage. The formation of a hurricane cannot be predicted but satellite photographs are used to identify hurricanes as they develop. The typical path of a hurricane in the northern hemisphere includes initial movement toward the west and northwest followed by a curving path that takes it northward and then toward the northeast. About 30% of hurricanes take more unusual paths that may include a loop.

The National Hurricane Center uses many different models to predict the future location of a hurricane. These are based on climatology, persistence, circulation, and dynamic or mathematical models.

Some notable hurricanes are Beulah for the number of tornadoes, Camille and Michael as category 5 intensities. Camille struck the Gulf Coast as a very devastating hurricane, then continued to produce heavy rainfall as it became an extratropical cyclone and flooded parts of the eastern United States. Camille had the strongest winds of any storm before it with gusts to 205 mph. Beulah generated more than one hundred tornadoes to surpass the previous record of 26.

The Galveston hurricane of 1900 killed more than 6,000 people making it the greatest natural disaster in the United States.

Four hurricanes have been seeded, several decades ago, with dry ice and silver iodide in an attempt to decrease their winds. This effort was largely unsuccessful and was terminated.

Hurricane watches and warnings are issued by the National Hurricane Center and relayed by the National Weather Service to the public. Appropriate preparations and reaction to hurricane warnings are important in saving lives from these storms.

Joe R. Eagleman

13

Floods and Drought

As the highway entered the canyon we noticed that it followed the small stream ahead. The stream looked slightly swollen but this was of little concern. Suddenly, as darkness closed in, our headlights met a wall of oncoming rushing water. The next thing we knew, we were struggling to get out of the van as it was carried along by the racing current. The headlights became useless as they sank beneath the muddy water; we were now fighting for our lives within the raging current. In almost total darkness, with sheer canyon walls on both sides, the two of us tried desperately to stay together clutching at any debris that might help keep us afloat.

As the current carried us against one of the walls of the canyon, we clawed at the smooth rock desperately searching for a way out of the torrent. As my hand touched a ledge, I grabbed Jim and we desperately struggled to drag ourselves out of the water. We found that the ledge was only large enough for one person and even then, it required effort to stay on it. We soon discovered a ledge of similar size just above us, and Jim proceeded to climb to it.

After more than an hour on the ledge, I felt something hit my arm and lodge there. As I picked it up I realized it was the radio transmitter we normally carried. Knowing that it could prove to be a very important piece of equipment in our circumstances, I made sure that I held on to it during the rest of the sleepless night. After an eternity, the morning sunlight began to light the canyon and we realized we were more than 30 ft. above the ground on the only ledges in the whole canyon wall within sight. As we finally contacted civilization through the transmitter we realized that no one was going to believe we had survived such a flood on these tiny perches.

Truth is sometimes stranger than fiction.

FLOODS

Few areas of the world are free from at least an occasional flood. Even the Desert Southwestern United States occasionally has a thunderstorm that produces very heavy localized runoff. Figure 13-1 shows a comparison of various areas of the United States troubled by floods. The North Atlantic region is most prone to flooding and major flooding occurs throughout the Missouri River Basin and other large basins.

Severe floods in the United States result from a number of different types of weather systems. The rainfall from hurricanes is one of the obvious causes. Hurricane Camille was described in the previous chapter as it did $1.4 billion damages and had more than 100 deaths from flooding

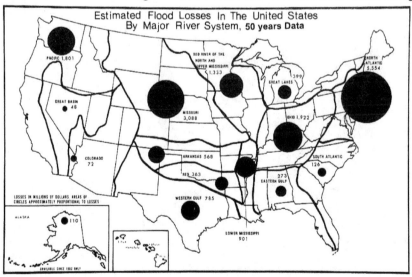

Figure 13-1. The flood losses in major river systems based on several-year records show considerable variation in distribution. The North Atlantic region, the Missouri River Basin, and the Pacific Coast all experience considerable flood losses. (Climatological Data, National Summary)

in New England. It is amazing that a second hurricane only 3 years later covered the same area with more than 100 deaths from floods once again in New England. Hurricane Agnes, in June of 1972, was responsible for one of the greatest economic flood losses in the history of the United States. Over 4 billion dollars' worth of damages were caused by Hurricane Agnes and 105 lives were lost mainly from floods in the East.

Another great economic flood loss developed the following year in 1973, when the entire Mississippi River system was flooded. Over one billion dollars' worth of damages were reported and 33 lives were lost.

Another of the worst damaging floods, disregarding those caused by dams breaking or hurricanes, was in July of 1951 when the Kansas and Missouri Rivers flooded. These caused over 900 million dollars' worth of damages and 28 lives were lost (Figure 13-2).

Other record breaking floods were those in Rapid City, South Dakota, in June of 1972 and the Big Thompson, Colorado, flood in July 1976, both resulted in more than 200 deaths.

Figure 13-2. This photograph taken on July 14, 1951 shows the severity of flooding in Kansas City. The foreground includes a residential section with the roofs of some of the houses still visible above the water. (Official US Navy Photograph)

CAUSES OF FLOODS

The primary cause of floods is extraordinarily heavy precipitation, but a smaller amount of precipitation may also produce flooding when it falls on ground that is already saturated. Precipitation from any given storm is partitioned into the amounts disposed of by infiltration into the soil, surface detention, evaporation and runoff. The amount of runoff into streams is, of course, the major factor in producing floods. The infiltration rate is related to soil characteristics. If the soil has a coarser texture, water soaks into it more rapidly than if it has a finer texture. Management practices such as cutting vegetation or overgrazing also affect the rate at which water will be absorbed into the soil.

Infiltration rates may vary from 2 in. (5 cm) per hour at the beginning of a storm to less than one half inch per hour as rainfall continues. Stream flow is composed of a certain amount of continuous base flow that represents the ground water flow into a stream. Surface runoff during a rainstorm adds to this base flow. Most streams or rivers have a definite floodplain (Figure 13-3). This is the low lying ground surrounding the borders of the river. The river may flood this lowland once per year, or once per decade, depending on the river channel and amount of precipitation. Above floodplains are broader steplike

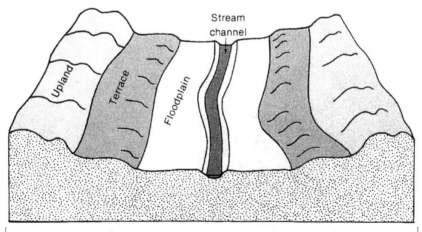

Figure 13-3. Rivers flow through a stream channel and are surrounded by a relatively flat area forming the floodplain. Outside the floodplain, terrace and upland regions exist. Flooding is normally confined to the floodplain with occasional flooding of terrace locations.

expanses of terrain called terraces. These may flood each decade or century.

The greatest observed precipitation rates are of interest. Over 1.23 in. of rain fell in one minute in Unionville, Maryland, and more than 12 in. of rain fell in only 42 minutes in Holt, Missouri. These extreme rainfall rates are much more than needed to produce substantial flooding.

The greatest economic flood loss in United States history was produced by the remnants of Hurricane Agnes in 1972, as previously stated. During the week preceding Hurricane Agnes, frontal activity brought soaking rain to the mid-Atlantic region, from New England to Virginia. Showers and thunderstorms dumped as much as 5 in. of rain on this area. In Central Pennsylvania 3 in. were common. Throughout much of New England, the soils were already on the way to saturation before Agnes arrived. Florida also had rainfall for two or three days prior to the passage of Hurricane Agnes. Floods developed along the Gulf as Hurricane Agnes formed in the Gulf and approached land. Among the

Figure 13-4. Wilkes-Barre streets during the Great Flood of 1972. (Times Leader Photograph)

highest total rainfall amounts were about 8 in. at Maple and Tallahassee, Florida. The flooding as Hurricane Agnes struck land was much less than after it was modified into a frontal cyclone. As it moved northward, and was modified when it interacted with the jetstream, as previously described for Hurricane Camille, it began to dump heavy rain into Virginia and much of New England. This produced severe flooding on the James and Appomattox River basins and along the Potomac and smaller rivers.

In the eastern half of the James River basin and western half of the Appomattox basin, flooding was the worst in history. Crest stages exceeded those of Hurricane Camille in 1969 and topped high water records dating back two hundred years. The average rainfall over the whole James River basin was 6 in. from the 19th to the 22nd of August 1972. New crest stage records were set at many cities along the James River. The river swamped a 200-block area of downtown Richmond in the worst flood in the city's history. The crest level set at the city locks on the 23rd of August topped the old mark set back in 1771.

The Potomac River began flooding on the 22nd of August. In downtown Washington, D.C., at Wisconsin Avenue, a flood crest of more than 12 ft. lasted for about 8 hours. In Pennsylvania, Agnes' heavy rains also fell on wet ground. Small streams began flooding first, followed by flooding of the major rivers. The Susquehanna River exceeded previous flood stage marks set in March of 1936 (Figure 13-4). Crests were generally more than 12 ft. above flood stage. At Wilkes-Barre, the water came over the levees completely flooding the city with a crest more than 18 ft. above flood stage. Thus, Hurricane Agnes produced major flooding, not as a hurricane, but in the form of an intense frontal cyclone.

REPEATED FRONTAL CYCLONES

A very different meteorological condition was responsible for the second greatest economic flood loss in United States history. The flooding of the Mississippi River in 1973 was caused by repeated frontal cyclone passage through the Central United States. The jetstream flow patterns are very influential in determining the amount of precipitation over large areas. Repeated frontal cyclone passage of both the longwave and trough type occurred over the Mississippi River basin from the 20th of March 1973 through the 9th of April 1973 (Figure 13-5). This

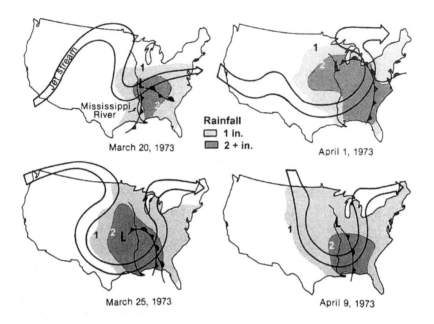

Figure 13-5. The Mississippi River flood of 1973 was produced as frontal cyclones repeatedly soaked the river basin during March and April.

brought widespread rain over the Central United States.

By the end of March, the Mississippi River was above flood level from Clinton, Iowa, to Donaldsonville, Louisiana, with some additional flooding at stations upstream as far north as Minnesota. Major flooding also occurred along the lower Missouri, Ohio and other tributary streams in the Mississippi Valley.

During April the severe overflow of the Mississippi River continued. This flooding was aggravated by additional periods of heavy rainfall from frontal cyclones during April. The Mississippi River rose to record high levels from Burlington, Iowa, to Cape Girardeau, Missouri (Figure 13-6). Millions of acres of rich farmland were inundated throughout the Mississippi Valley. At Memphis, Tennessee, the Mississippi was over flood stage for 63 days, more than that of the historic 1927 flood, and the river was above flood stage for an even longer 107 days at upstream, Cairo,

Figure 13-6. This photograph taken April 30, 1973, includes the main street of West Alton, Missouri. This town located about 6 mi. from the confluence of the Missouri and Mississippi Rivers became the joining point when a number of levees broke on both rivers. There was considerable concern at this time that the rivers may have cut a new channel through parts of the town. (Official US Coast Guard Photograph)

Illinois, tens of thousands of people were evacuated from their homes. Entire populations of some communities were evacuated; damage to roads, buildings, and bridges was extensive. Damages were estimated to be one billion in today's dollars.

STATIONARY FRONT

Another flood producing meteorological condition is established as a well-defined front stalls and becomes stationary. This happens when either a warm or cold front becomes parallel to the axis of the jetstream. Thus, the upper level winds are along the front rather than across it to move it along. If the stationary front has persistent south winds

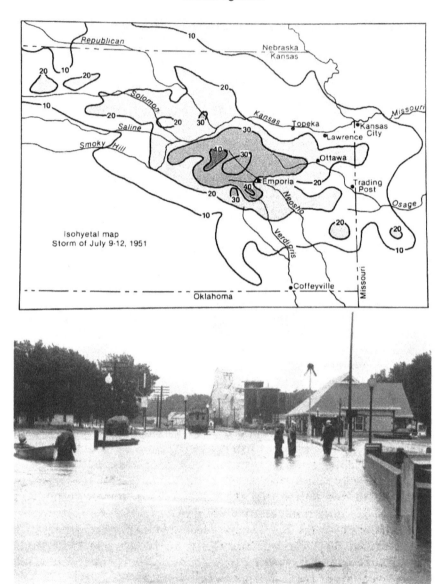

Figure 13-7. The top map shows the number of centimeters of water that fell over eastern Kansas on July 9-12, 1951. Major flooding occurred in all low-lying areas. The bottom photograph shows north Lawrence, Kansas on July 14, 1951. (Official US Navy Photograph)

southward from it and north winds northward, the mechanism is present for producing precipitation in the same locality for an extended period of time. This was the meteorological condition that produced the flood responsible for another extremely large economic flood loss-the Great Flood of the Kansas-Missouri rivers in 1951.

During the 72-hour period, July 9 to 12th, excessive rainfall fell over much of the eastern half of Kansas (Figure 13-7). Much of the Kansas River Basin received more than 25 cm (9.8 in.) of water on top of previously wet soil from the 25 cm or more of rainfall during June that fell over all of northeastern Kansas. This caused severe flooding of cities along the Kansas River, including Kansas City, Missouri and Lawrence, Kansas.

INDIVIDUAL THUNDERSTORMS

It is not too unusual for a single thunderstorm to produce flashfloods that result in loss of lives. Thunderstorms with a capacity to cause flood conditions can occur throughout almost any of the tropical or midlatitude locations. Even the Sahara Desert is subjected to an occasional thunderstorm. Some of the greatest killer floods have

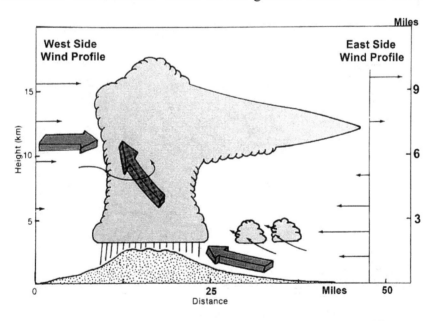

Figure 13-8. Flash floods in mountainous terrain are caused by very different weather patterns than occur over the plains. A thunderstorm may produce heavy rainfall in a localized area when the upper level winds are very light and opposite to the low-level air currents.

resulted from the rainfall of thunderstorms.

The weather patterns for developing flood producing thunderstorms in mountainous areas are quite different from those that produce thunderstorms with heavy rain in the Great Plains. If the jetstream is absent and the upper level winds are light, weak southeasterly surface winds help set the stage for heavy rainfall in mountainous areas (Figure 13-8). Such a pattern developed over Colorado on July 31, 1976. A flash flood was produced in the Big Thompson River Canyon 40 miles north of Denver, Colorado. The Big Thompson River ordinarily is less than 5 ft. deep but at 6 pm on July 31st, it started to rain and within the next six hours thunderstorms had produced 14 in. of rainfall. By 9 pm a wall of water about 30 ft. deep was rushing down the Big Thompson Canyon taking everything with it. It caused $9 billion (today's dollars) worth of damage and killed more than 200 people. The opening paragraphs of this chapter were based on the accounts of two survivors of this flood.

The surface and upper atmospheric weather charts for July 31 1976, show a surface front located over Colorado. The 500 mb winds were very light with a ridge in the jetstream flow patterns and a

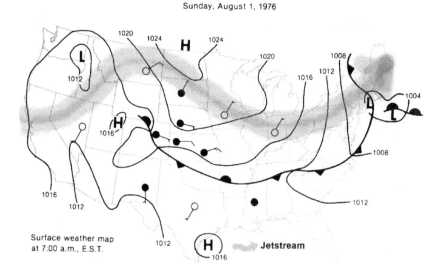

Figure 13-9. The surface map and location of the jetstream corresponding to the Big Thompson flood in Colorado shows light surface winds from an easterly direction. Very light upper level winds, because of the northward displacement and ridge in the jetstream, are from the west. Such weather patterns are typical of those producing flash floods in mountainous areas.

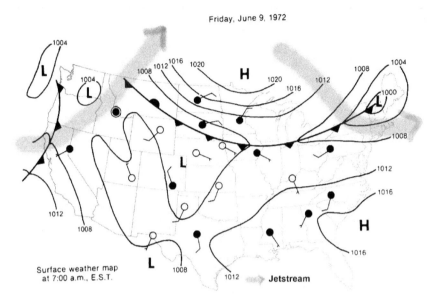

Figure 13-10. The surface weather map for the Rapid City, South Dakota flood shows light surface winds and the presence of a stationary front with light upper-level winds because of a ridge and northward displacement of the jetstream. These conditions are typical of those producing flooding in mountainous terrain.

general high-pressure area located over the Great Plains (Figure 13-9). Thus, thunderstorms building over the Big Thompson Canyon were not moved along by the upper airflow and were fed by the low-level winds as was shown in Figure 13-8 to give a continued downpour for several hours. Note that the actual winds around this stationary storm are similar to the relative winds around moving thunderstorms in the Great Plains. Although the rain was very localized, and the heavy rain extended over a distance less than 15 mi., it was centered over the mountainous terrain that fed the Big Thompson River.

A similarly devastating flood occurred in Rapid City, South Dakota on June 9, 1972. The surface and jetstream characteristics were amazingly similar to those causing the Big Thompson flood (Figure 13-10). A cold front pushed through South Dakota on June 9th beneath a ridge in the upper airflow pattern with very light winds. In 1972, South Dakota's operational weather modification program was just getting into full swing. A rain gage network had been set up at strategic locations to measure the effect of seeding clouds in the Rapid City area. More than 550 lbs. of salt were used for seeding

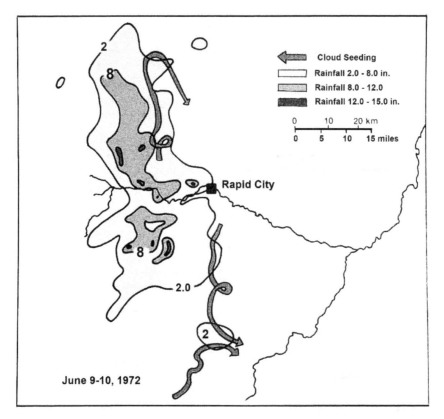

Figure 13-11. Cloud seeding was conducted only a few hours prior to the Rapid City flood. Even though all the factors were present to produce the flood from the natural weather factors. This seeding operation had a detrimental effect on weather modification in South Dakota (After Armand, Report on Rapid City Flood of 9 June 1972.

clouds near Rapid City between 3 and 5 pm on June 9th. Heavy rainfall began about 5 pm and thunderstorms produced more than 10 in. of water in some localities near Rapid City (Figure 13-11). All the seeded area received more than 2 in. of rainfall.

At the 40-year anniversary of the fateful flood, survivors remarked on the horrific events. Rita, who was 20 at the time, described the scene, "There were so many people in trees and screaming and crying and the sparks were flying from electric wires, houses were on fire, it was just — it was hell," she says. Rita was seven months pregnant at the time of the tragedy. She describes her fears as, "I wouldn't wish that upon nobody,"

Figure 13-12. Some of the battered cars in Rapid City. (Photograph, Rapid City Journal)

she says. "That's a nightmare and I have to think that you're going to die in water and your mom is going to go with you and you're trying to do your best to keep your mom alive." Rita and her mother were swept against a building and thankfully rescued. There were others with the same nightmares while others were worse. Good Samaritans, like Alex were left to clean up the mess and search for the less fortunate.

The flood caused a tremendous amount of damage. Flood waters displaced large rocks, trees, trailers, and vehicles, and carried homes away. In Rapid City the flood resulted in the deaths of 238 people and over 3,000 being injured. More than 3000 homes were destroyed or damaged and 5,000 cars were demolished (Figure 13-12). The damage in Rapid City totaled $66 million in 1972-dollar value. As for Keystone, eight people were killed and much of the town was washed away. The total cost of the 1972 Black Hills flood totaled $165 million, including infrastructure and utilities. In todays dollars that would be just over one billion dollars.

Although all the natural factors for producing flood conditions were present in South Dakota on June 9, 1972, the fact that weather modification was attempted within the area resulted in loss of public support for the operational weather modification program in South Dakota and the program was canceled.

SNOWMELT AND OTHER FLOOD CAUSES

Various other weather factors that may be involved in producing flash floods include a slow-moving frontal cyclone. If a weather system moves very slowly then it affects the same area for a longer period of time; thus, the potential for flood conditions is increased.

Rapid increases in temperature may be responsible for causing floods during the spring if a heavy snow cover is on the ground. Rainfall on top of a snow cover is also a frequent flood producing condition. Frozen ground may also be a factor during the spring months. If the warming in the spring occurs rather rapidly, and the ground is still frozen underneath, the water does not infiltrate into the ground and therefore a greater amount goes into the runoff segment of the hydrologic cycle. Thus, various combinations of snow, temperature, and rainfall may be responsible for flood conditions.

FLOOD WARNINGS

The National Weather Service operates a flash flood warning service. **Flash flood watches** and **warnings** utilize such information as the quantitative precipitation forecast. Quantitative precipitation forecasts give the expected amount of precipitation within the next 24 hrs. These are used along with such other information as satellite photographs, radar, and surface observation networks. River stations may also monitor critical water levels for several miles upstream from cities.

Sophisticated systems have been developed for automatic warnings in many flood prone regions. The concept of the Integrated Flood Observing and Warning System has been developed extensively since the creation of the National Flash Flood Program Development Plan in 1978. The goals of the program are to substantially reduce the annual loss of life from flash floods, reduce property damage, and reduce disruption of commerce and human activities. When a flash flood watch or warning is issued by the National Weather Service, it is transmitted to local news media for appropriate community action.

DROUGHT

The importance of water is rapidly emphasized if it becomes unavailable. Drought is a deficiency of water and can be best defined in terms of departure from normal precipitation. A county or city

Figure 13-13. The dust within a dust storm such as this one of the 1930s became so dense that it was a threat to life and property. (Courtesy of the Library of Congress)

becomes accustomed to the amount of rainfall it normally receives, and all water related activities are conducted accordingly. When departures from normal amounts occur, they may create severe problems. The 1930s are well known for dust storms (Figure 13-13). More recently southern California experienced below normal amounts of rainfall during the winter months of 1976 and 77. The drought became so severe that by early 1977 water was rationed because of the shortage. A drought in the early 1960s in the northeastern United States was the worst in 160 years. It brought widespread unemployment and critical water shortages throughout most of the heavily populated Eastern United States. In New York City, restaurants no longer offered a glass of water. The use of water by individuals was also restricted; people were not allowed to water lawns or wash cars, even the water hydrants were locked tight to prevent access to this water source. By 1965, the reservoirs of the city of New York were down to 25% of their capacity and the reservoirs of Washington, D.C. forced residents there to face the real possibility of water rationing.

Many different types of losses result from drought. Obvious losses are agricultural crops, trees, and grass (Figure 13-14), but many additional kinds of losses are equally important. These include business losses such as lost income from investments, lost revenues to cities because of lack of water. It is, in fact, difficult to determine the exact

Figure 13-14. Extreme soil erosion can result from a combination of dry soil, lack of vegetation, and strong winds. A South Dakota wheat farm is shown above as observed in August 1970. (Courtesy of USDA Soil Conservation Service)

economic significance of drought because of its nature. For example, the onset of drought in a distant country may affect the price of bread on the other side of the world. Drought effects are intensified as civilization expands into deserts as far as possible. Continual drought exists in deserts, but we call this condition aridity. However, on the fringes of deserts the water supply is marginal and is therefore adequate in some years and absent or deficient in others. Thus, normal variations in rainfall have very disastrous results in semiarid regions.

DROUGHT PRODUCING WEATHER PATTERNS

Very dry and hot weather conditions are generated in the Central and Eastern United States by a ridge in the jet stream. Such a flow pattern was present during the summer of 1965 as extreme dry conditions were experienced in the Eastern United States. A ridge also became stationary over the Central United States in 1976 and 1980 to produce extremely dry weather. If the jet stream is displaced farther northward,

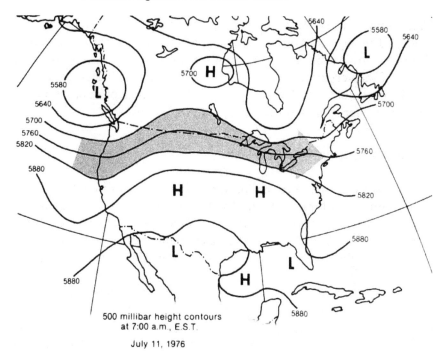

Figure 13-15. Drought is typically produced from a ridge in the upper airflow pattern, such as that shown above for July 11, 1976. Surface high pressure systems are persistent beneath an upper level ridge and also contribute to continued dry weather.

this leaves very light upper level winds over the United States with resulting hotter and dryer weather (Figure 13-15).

Along the West Coast, drought is frequently influenced by the position of the Hawaiian high-pressure area. This semipermanent high-pressure region normally shifts southward during the winter to allow frontal activity to bring rain to the West Coast. If, on the other hand, this high-pressure area remains further northward during the winter months the result is dryer than normal weather. Since much of the West Coast receives almost all of its rainfall during the winter months, the position of the Hawaiian High is crucial for their water supply.

In tropical regions the Intertropical Convergence Zone is important in determining which areas will receive rainfall. As the Intertropical Convergence Zone, shown previously for hurricane formation, moves with the seasons, the converging air between the northeasterly Tradewinds of the northern hemisphere and the southeasterly

Tradewinds of the southern hemisphere cause deep convective rain clouds. The movement of the Intertropical Convergence Zone varies from year to year bringing variability of rainfall amounts. A primary factor of the very devastating Sahelian drought of the early 1970s was the displacement of the Intertropical Convergence Zone about 2° latitude further southward than normal across Africa.

AGRICULTURAL DROUGHT

The effect of dry weather on agriculture is determined not only by the amount of rainfall but by the amount of water lost to the atmosphere. The maximum water loss rate is called the potential evapotranspiration. This is defined as the amount of water lost from actively growing crops with an abundant supply of moisture in the soil. The potential evapotranspiration rate is determined primarily by meteorological variables such as temperature, relative humidity, winds, and radiation, with the greatest influence coming from the temperature and humidity.

The actual evapotranspiration at any time is governed by still another factor- the amount of soil moisture. Since abundant water is often not available for plant use, the actual evapotranspiration rate is frequently less than the potential

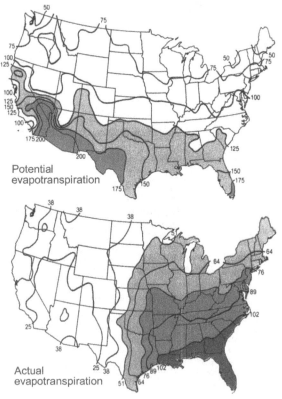

Figure 13-16. The potential and actual evapotranspiration (cm) shows considerable variation throughout the United States.

rate. The distributions of actual and potential evapotranspiration are shown in Figure 13-16. The potential evapotranspiration is more than 79 in. (200 cm) per year in much of the desert southwestern United States. Across the northern United States, it is only about 30 in. (75 cm). The actual evapotranspiration rate is much less than the potential rate in those areas where precipitation is small, thus the Desert Southwest loses less than 10 in. (25 cm) per year, while the Gulf Coast states lose more than 39 in. (100 cm) per year.

Water balance diagrams reveal much about the climate of a location. These can be made by plotting the monthly values of precipitation, and

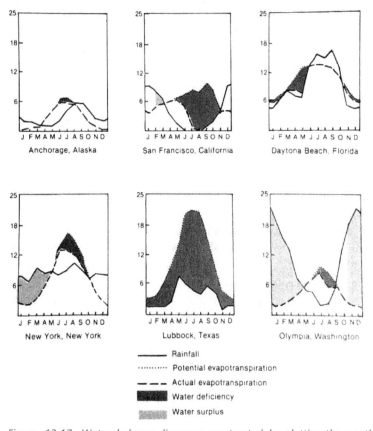

Figure 13-17. Water balance diagrams constructed by plotting the monthly rainfall, potential evapotranspiration, and actual evapotranspiration rates (shown above in cm) are useful in showing the moisture characteristics of a region. (After Eagleman, The Visualization of Climate, D.C. Heath and Co., Lexington, Mass., 1976)

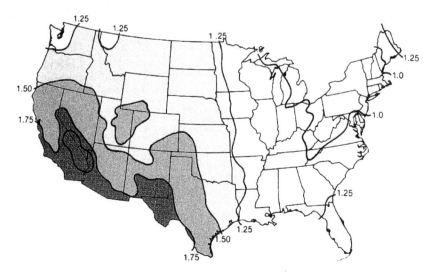

Figure 13-18. The variability of rainfall expressed by a statistic similar to the coefficient of variation shows that rainfall is most variable per inch in the Southwestern United States. It is least variable per inch of rainfall in the Northeastern United States (After Eagleman, The Visualization of Climate, D.C. Heath and Co., Lexington, Mass., 1976.)

potential and actual evapotranspiration. Such diagrams are useful means of describing the overall moisture conditions for a region. Specific water balance diagrams for several locations are shown in Figure 13-17. If the potential evapotranspiration rate exceeds the amount of water available from precipitation and soil moisture a deficiency of moisture exists. The magnitude and duration of moisture deficits are of major importance to agriculture. If soil moisture deficits exist for an extended period of time, they may have a devastating effect on agricultural production.

The southwestern United States has the greatest accumulated deficiency of moisture. In addition to the amount of rainfall, the variability is extremely important in semiarid regions. The variability of rainfall in the United States as shown in Figure 13-18 is greater in dryer areas than in wetter areas. The amount of precipitation also varies with time. Some indication exists for a 20 year cycle in rainfall variability. The 1930s were extremely dry in the Central United States; drought conditions prevailed in the 1950s with some dry years in the 1970s. The cycle was less obvious in the 1990's and 2010's.

Continued drought can change the landscape (Figure 13-19). The

Figure 13-19. The extreme dust storms of the 1930s completely changed the landscape in some locations. (Courtesy of the Library of Congress)

dust storms of the 1930s covered fences with mounds of dust; a dust storm 300 mi. in diameter can carry 100 million tons of dust.

Dust storms are caused by a combination of lack of precipitation and little or no soil cover. Dust storms are not limited to the 1930s; a major dust storm occurred on February 23, 1977. As a strong longwave cyclone traveled eastward it picked up dust from Colorado, New Mexico, and Texas and carried it along south of the longwave cyclone center across the southern United States and into the Atlantic. The dust concentration was large enough to reduce the visibility to almost darkness in the middle of the day. The dust also showed up on satellite photographs (Figure 13-20).

Recent dust storms have not matched those of the 1930s, however. From 1935 to 1937, unusually strong winds combined with the extended drought to produce great clouds of blowing top soil which obscured the sun over much of New Mexico, Colorado, Oklahoma, Kansas, and Texas. The grim experiences of the 1930s revealed that the semiarid lands of the Great Plains could not be farmed by the same methods as the farmlands in the more humid East. Many farming practices have changed accordingly. New machinery is used that leaves the soil surface less disturbed, terraced farming is practiced, and more crop residues are left on the ground to help stabilize the

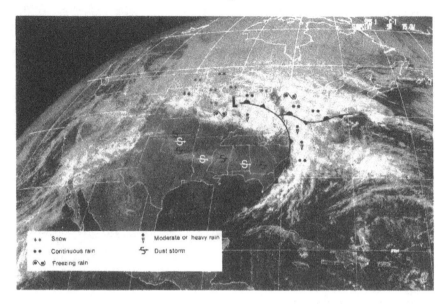

Figure 13-20. A major dust storm is shown in this satellite photograph taken February 24, 1977. The light area extending through Oklahoma, Southern Arkansas, and other southern states is caused by extremely dense, blowing dust.

soil and help it to absorb and retain more moisture.

In the early 1950s, from 1950 through 1956, drought again returned to Texas, Oklahoma, New Mexico, Arizona, Kansas, Missouri, Colorado, and Nevada. An analysis of tree rings, at the University of Arizona, brought the conclusion that the drought of the 1950s was the worst to afflict the Desert Southwest in 700 years. More than half the farmers in much of the afflicted area had to find jobs off the farm. The national disaster of the 1930s was not repeated, however, because the national economy was strong and growing with plenty of jobs in towns and factories.

Satellite imagery has shown how dust storms on another continent can affect regions thousands of miles away. Figure 13-21 shows dust from the Sahara desert traveling to Mexico where it may eventually make its way to the southern United States.

Figure 13-21. This "true-color" composite image of the Saharan Dust plume was captured by the VIIRS instrument aboard NASA/NOAA's Suomi NPP satellite on June 24, 2020. (NASA/NOAA Photograph)

SUMMARY

A nation's water supply is probably its most important asset; however, it does not arrive from the sky in an evenly distributed and measured quantity. Floods are caused by several different specific meteorological factors. Repeated frontal cyclones were responsible for the flooding of the Mississippi River in 1973. The ground became saturated and as the rains continued more water ran into the streams and rivers creating severe flood conditions along the Mississippi River.

If a front becomes stationary, it can also cause flooding. This was responsible for a very damaging flood in 1951 in Kansas and Missouri. Flash floods may result from individual thunderstorms. A single thunderstorm that drops several inches of water within a few hours may create flood conditions. This occurred in the Big Thompson Canyon in July of 1976, and also in Rapid City, South Dakota, in 1972. Several other factors including snow melt may cause or contribute to the generation of floods.

Flash flood watches and warnings are prepared by the National Weather Service. Since flash floods are caused by thunderstorms that are localized events their forecasting is somewhat comparable to forecasting tornadoes. It is important to know that flash floods can occur suddenly and are, in general, smaller in scale than the distance between reporting weather stations. However, the Nation Weather Service has established automatic remote warning systems, that measure water levels, for many locations that are prone to flooding.

At the other extreme in moisture supply, drought conditions are a major problem in many parts of the world. Drought conditions are generally produced by specific upper-air flow patterns in the atmosphere. A ridge in the jetstream developed over the Eastern United States in the 1960s was responsible for drought conditions there. A ridge in the jetstream that persists for several weeks decreases the chance for rainfall.

The effect of drought may be disastrous on agriculture, since vegetation requires plenty of water for maximum yield. If the evapotranspiration rate requires more water than is available from precipitation and moisture stored in the soil, crops suffer and yields are reduced. Severe crop losses and dust storms formed in the 1930s. These were generated by a combination of lack of rainfall, poor farming practices, and strong winds. Such dust storms still occur but are, in general, less frequent and less intense.

14

Unusual Storms and Weather Patterns

As I watched the whirlwind travel across the barren field, I wondered about the strength of the winds inside and how long it would last. As I watched it make its way toward a farmhouse, I noticed that a dog was approaching the swirling air made visible by dust picked up from the dry ground. I watched in amazement as the dust devil surrounded the dog, lifted it into the air and took it up to a height of about 20 ft. before it reached the outer part of the vortex where it was lowered down to the ground, apparently without harm. As the dust devil moved closer to the mountains it appeared to intensify, and as it approached the highway it left little doubt that it had become a strong vortex as it turned a car around and set it in the ditch. The dust whirl that had started as a very small atmospheric disturbance was now so strong that human safety demanded that it be avoided.

DUST DEVILS

Dust devils are, in general, the smallest type of storm if they can be called storms at all. If you have ever traveled across the Desert Southwest in the summer, you have probably noticed dust devils or whirlwinds in action (Figure 14-1). They can develop from rising, unstable air by winds flowing around obstacles or by vortex twinning. Vortex twinning may occur as a vortex with a horizontal axis develops because of increasing wind speeds with height. As it hits obstacles,

Figure 14-1. A large dust devil in progress across the desert in Arizona is shown here. Such dust devils may vary greatly in strength. (Courtesy of Sherwood Idso.)

the vortex may be tilted to the vertical to form two dust devils where the ends touch the ground; thus, the name vortex twinning (Figure 14-2). Warm surface air is important in dust devil generation just as in the generation of hurricanes. A further similarity exists in that both consist

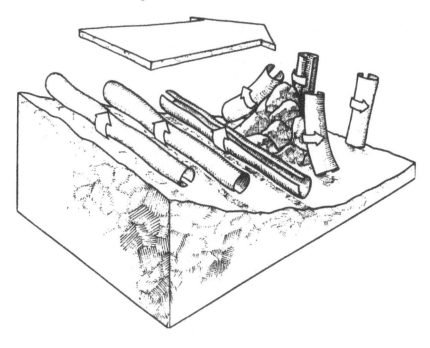

Figure 14-2. The process of vortex twinning occurs as horizontal circulation caused by wind shear is tilted into the vertical by topographic barriers. (After Sherwood Idso, Weatherwise, 1976.)

of spiraling air around a downdraft core. The dust devil, however, in contrast to the hurricane, revolves in either the anticyclonic or cyclonic direction. It is not generated by the larger circulation and is not affected by the force caused by the earth's rotation since it is so small. A major generating mechanism is surface heating, while the downdraft core simply replaces air that is rising in the surrounding vortex.

MOUNTAINADO

A slightly larger storm called a **mountainado** forms in some mountainous regions. Boulder, Colorado, has an unusually high frequency of winds greater than 70 mph. Damage incurred cannot ordinarily be explained by straight winds. Houses along a particular street, for example, will have windows blown out, roof shingles lifted off with the nails pulled out, indicating suction similar to a tornado. Such damage appears to be associated with vertically oriented vortices containing winds of about 90 mph.

Several such examples have been recorded in Boulder. On January 18, 1971, the wind damage pattern was such that it led to the conclusion that high velocity winds were accompanied by strong twisting action. On the 21st, according to an eyewitness account, the roof of a recently completed house was lifted 45 ft. into the air before it fell back to earth. The contractor who was working on a house next door described the wind as a circular motion dust devil.

A similar mountainado was described several years earlier by a man in Boulder who with his wife watched a strong whirlwind lift their 18 x 30 ft. shed about 30 ft. in the air to destroy it completely as it spiraled around in the air. Again, in 1975, on the 30th of November, during a time of relatively strong winds with 90 mph gusts, vortices were observed to develop and move away from the mountains over snow covered ground. The mountainadoes were about 100 ft. in diameter and from the snow they picked up to make them visible, it was determined that they were about 100 ft. in height. The horizontal winds in them were 45-95 mph.

Since mountainadoes are generated in winter as well as summer, their formation mechanism is different from that of dust devils. The dynamics of airflow around an obstacle, such as the foothills, tend to initiate double vortices behind the barrier in the same way that the double vortex is initiated behind a strong updraft that blocks the airflow around a thunderstorm. Flow over the top of the barrier

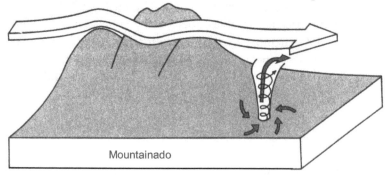

Figure 14-3. A vortex is intensified as air is extracted through its central core. Thus, a small vortex may be intensified until it has sufficient strength to damage houses and other properties. Mountainadoes are formed by airflow over topographic barriers interacting with a vortex that may extend to the surface.

Figure 14-4. Vortices may be formed by forest fires and other heat sources. Two vortices are formed here by the Meteortron in France. (Photograph courtesy of Christopher R. Church.)

intensifies the vortex just as strong upper air flow intensifies a tornado (Figure 14-3). Thus, we have the same mechanism on a small scale that generates a tornado on the large scale, i.e., horizontally circulating air with strong wind over it to intensify the updraft and strengthen the circular motion of the vortex.

THERMAL VORTICES

Thermal vortices are observed to form from thermal plumes from such sources as forest fires, volcanoes and industrial operations. An experiment set up in France by the Centre de Recherches Atmospheriques Henri Dessens uses an array of fuel oil burners, called the Meteortron, to supply 1000 MW of heat. The resulting thermal plume regularly produces vortices. One method of vortex generation occurred as the rising plume interacted with moderate winds to form a double vortex structure (Figure 14-4). It is interesting to speculate whether the formation of these two counterrotating vortices share

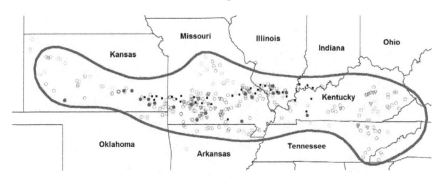

Figure 14-5. The Super Derecho outbreak of May 8, 2009. (SPC, NOAA)

anything in common with the formation of the double vortex thunderstorm on a large scale.

DERECHO

A **derecho** is a widespread, long-lived, straight-line wind storm that is associated with a fast-moving group of severe thunderstorms known as a mesoscale convective system and potentially rivaling hurricanic and tornadic forces. A common place for them to occur is in the warm air sector of a frontal cyclone. They bear some resemblance to the gust front produced by the down-rush of cold air in the leading part of a thunderstorm. They are frequently associated with thunderstorms that produce a bow echo.

The May 8, 2009 Super Derecho outbreak (Figure 14-5) was one of the most intense and unusual ever observed. The storms produced significant and often continuous damage over a broad swath from the High Plains of western Kansas to the foothills of the Appalachians in eastern Kentucky. Multiple wind gusts exceeding 70 mph with gusts over 90 mph were measured. Numerous bow echoes and a well-defined larger bow and some tornadoes were seen. Some flash flooding also occurred.

El Niño and La Niña

The term **El Niño** refers to the large scale ocean and atmosphere interaction linked to a periodic warming in sea surface temperatures across the central and east-central Equatorial Pacific. An El

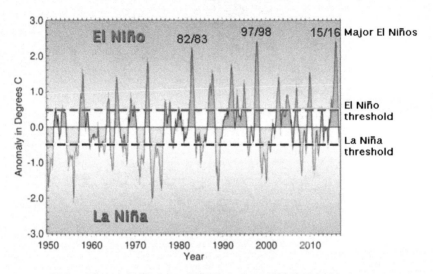

Figure 14-6. Cycles of El Nino and La Nina. (NOAA)

Niño condition occurs when surface water in the equatorial Pacific becomes warmer than average and east winds blow weaker than normal.

The opposite condition is called **La Niña**. During this phase, the water is cooler than normal and the east winds are stronger. El Niños typically occur every 3 to 5 years. In the Southern United States, during the fall through spring, El Niño usually causes increased rainfall and sometimes flooding.

La Niña, however, usually causes drier weather in the South, but the Northwest tends to be colder and wetter than average. Even though El Niño occurs in the Pacific Ocean, it often reduces the number of hurricanes that form in the Atlantic Ocean. Conversely, La Niña events tend to be related to an increase in the number of Atlantic hurricanes. Regular cycles of these events seem to occur as shown in Figure 14-6. But the cycles are not regular enough to allow their prediction for exact future years.

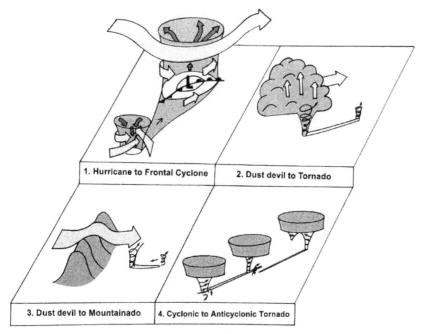

Figure 14-7 Storm metamorphosis occurs as hurricanes become frontal cyclones, dust devils become tornadoes or mountainadoes, and cyclonic tornadoes are converted to anticyclonic tornadoes.

STORM METAMORPHOSIS

Some interesting transitions in the nature of storms occur that we might call storm metamorphosis. Each of the different individual atmospheric storms has its own unique structure depending on the formation mechanism. But in some cases, it is possible for one type of storm to transform into a different type and take on an entirely different structure (Figure 14-7). The hurricane is an example already noted since it may become a frontal cyclone. In order to do this, the structure has to be changed from a downdraft in the center of the hurricane to an updraft over a larger area. The size of the whole storm must increase along with a reduction in intensity. Frequently, such a metamorphosis produces a frontal cyclone that is much stronger than average as some of the characteristics of the hurricane are retained. Good examples of such storms were the frontal cyclones produced by the remnants of Hurricanes Agnes and Camille that have already been covered as they produced very damaging floods.

Another example of storm metamorphosis is the dust devil

transformed into a tornado. The dust devil is ordinarily very weak but it can connect with a fairly small developing cumulus cloud overhead to become a tornado. A developing cumulus cloud is composed primarily of a large updraft without the more organized structure of the large thunderstorm that typically produces tornadoes. However, the updraft can strengthen a vortex beneath the cloud into a tornado even though it could not generate a tornado without the preexistence of the vortex in its initial form as a dust devil. A tornado generated in this way is called an **eddy tornado**.

Dust devils may also become mountainados. In this case airflow over a topographic barrier interacts with a dust devil on the leeward side to transform it into a mountainado.

The transformation of a cyclonic tornado to an anticyclonic tornado is a less complete form of storm metamorphosis. Such transformations have been observed beneath a single thunderstorm. The mechanism for the transformation is probably similar to the laboratory observations that have shown that small anticyclonic vortices tend to form around a cyclonic vortex because of wind shear. As the cyclonic tornado decays, an anticyclonic vortex may connect with the updraft to

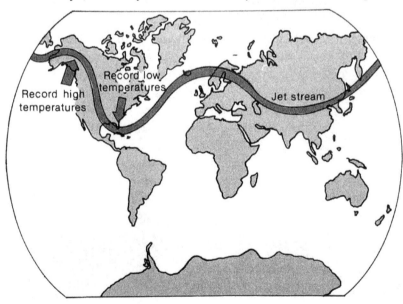

Figure 14-8. Unusual synoptic patterns such as large persistent meanders in the jet stream may cause record breaking high temperature in Alaska while temperatures may dip below freezing in Miami, Florida.

become the tornado supported by the thunderstorm. Such transformations-and anticyclonic tornadoes in general-are rather uncommon.

UNUSUAL WEATHER PATTERNS

Weather may also be unusual because of unusual airflow patterns. The meandering nature of the jet stream with warm air to the south and cold air to the north can produce record breaking high temperatures in one location, Anchorage, Alaska, for example, while another, such as Miami, Florida, is subjected to unusual cold waves (Figure 14-8). Since large meanders in the jet stream can persist for several weeks, such unusual temperature anomalies can also persist.

UNUSUAL COMBINATIONS

Other weather systems are unusual because of combinations of various factors. A good example is the **haboob** (Figure 14-9). This is an Arabic word from the Sudan meaning "to blow." Desert regions are affected by this particular storm which is unusual because it consists of a very dense wall of dust which may reduce visibility to less than 10 ft. The sudden onset of a haboob was one of the factors that contributed to the failure of the American effort to rescue hostages

Figure 14-9. The haboob is a very dense dust storm that is common in Arizona. This one was in Phoenix, Arizona. (Photograph by Sherwood Idso)

from Iran in 1980.

Phoenix, Arizona, often experiences these storms. They are caused by winds greater than about 60 mph that whip up loose dust from the desert. After the wind is laden with dust it causes damage by sandblasting. This removes paint, and kills trees and people. On July 16, 1971, a thunderstorm that caused a haboob moved out of the Santa Cruz river valley southeast of Tucson toward Phoenix at about 30 mph. The temperature dropped by 15°F and the relative humidity rose from 33% to 74%. Nine people were killed and 27 injured on highways because of reduced visibility. Phoenix has an average of 2 of these storms every year with as many as 12 in a single year.

The haboob is caused by the gust front of a thunderstorm which combines with the dust of the desert to produce the haboob. The gust front forms from the downdraft within the thunderstorm and spreads out ahead of it as previously described. Wind speeds of 60 mph are common. If rain occurs with the thunderstorm it is behind the haboob.

MISCELLANEOUS UNUSUAL WEATHER

In southern California a warm dry **Santa Ana wind** similar to the chinook wind off the eastern slope of the Rockies blows over the coastal

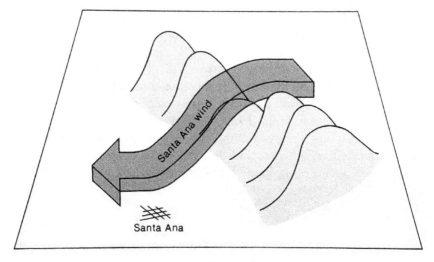

Figure 14-10. Strong winds are occasionally experienced at Santa Ana, California, as the pressure distribution forces air through particular mountain passes. As this air descends it is heated by compression and becomes a warm dry wind called the Santa Ana wind.

mountain ranges (Figure 14-10). When the pressure gradient favors east winds the flow of air through the mountain passes sometimes becomes concentrated and reaches speeds of 90 mph. The descending air is warmed adiabatically and may cause a rapid increase in temperature along with a rise in fire hazard since the relative humidity of the air also decreases rapidly as the air heats. These winds are called Santa Ana winds after a particular city near Los Angeles where they are unusually strong.

Other unusual types of weather are interesting, such as colored rainfall. The **red rains** of France and **blue rains** of Greece are caused by a combination of dust particles which serve as condensation nuclei within a thunderstorm. If the dust particles are colored, the rain droplets may appear colored. In some desert areas dust particles are red because of iron content. Copper minerals cause blue or green rain. **Yellow rains** in the Central United States fall from clouds that have picked up yellow-red dust from Oklahoma or Texas.

The weather may be unusual in a variety of other ways. On March 26, 1895, a thunderstorm produced a tornado that moved across Albany, New York, with accompanying snow instead of rain. Two years earlier, on December 16th, a tornado moved through Oswego, New York, accompanied by sleet and ice pellets. It is apparently never too cold to snow. At Fort Yellowstone, Wyoming, records indicate that 3 in. of snow fell on the 2nd of February 1899 when the maximum temperature did not get above -18°F all day. In contrast it snowed on the 4th of July in 1879 in Portland, Maine.

Thunderstorms that produce snow with associated thunder and lightning are unusual, but it is even more unusual for them to also produce St. Elmo's Fire. However, this was observed at Fort Huron, Michigan, on March 25th, 1930. During a fall of heavy wet snow, St. Elmo's Fire was observed at the end of the arrow of the wind vane and on the axis of the vane and the tail. It made hissing sounds loud enough to be heard 60 ft. below the wind vane.

CLIMATE CHANGE

A logical question to ask is how climate change affects severe and unusual weather. Trillions of dollars have been spent on research on climate change. Complex investigations using elaborate computer models are not able to include and successfully consider all of the many variables that control our climate. Figure 14-11 shows a few of the many variables

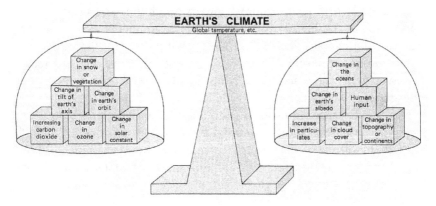

Figure 14-11. Our climate is determined by a large number of interrelated variables. A number of them that affect the climate scale are shown here.

that would have to be considered if we wanted an accurate answer concerning climate modification. Attention has been focused on the increasing carbon dioxide levels due to consumption of fossil fuels. Little attention is given to the fact that an increase in cirrus clouds of less than 1% would offset the amount of increased carbon dioxide for more than a decade.

The primary climatic control is the incoming and outgoing radiation balance and even the amount of incoming solar radiation can change over time due to a number of factors. Also, the amount of radiation reaching the earth is affected by such variables as cloud cover and particulates in the atmosphere. Suppose that the average global temperature increased by 1°. We would expect increased evaporation from the oceans and vegetation and this would lead to additional cloud cover. This could be enough to offset the increased temperature and return the global average back to its previous level. This is just an example of the codependence of only two of the many variables that would have to be known in order to predict the future state of our climate.

One of the most obvious means of seeing how much our climate is currently changing is to look at measurements of annual snow and ice cover in polar regions. This should be a tangible manifestation of climate change because the melting of snow and ice is very sensitive to the temperature. Very accurate measurements of the coverage of snow and ice have been obtained from satellites since 1966. These measurements could be expected to be a better measure of climate change than looking at the temperature record. Temperature records through time are confused by the growth of urban areas around where measurements are made. For example, the average increase in temperature in a city that grows to a population of 1 million is 7°F simply from the city itself. For that reason, the coverage of snow and ice in polar regions would seem to be a better indicator of climate change.

The satellite measurements of snow and ice are shown in Figure 14-12. These measurements show only a slight decrease in snow coverage in the northern hemisphere and a slight increase in sea ice in the southern hemisphere. These measurements do not show the major climate changes that some people have expected to see.

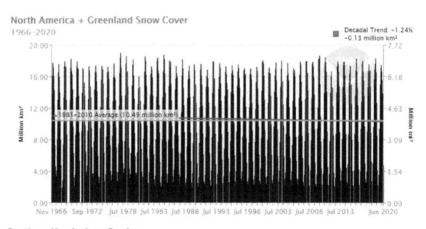

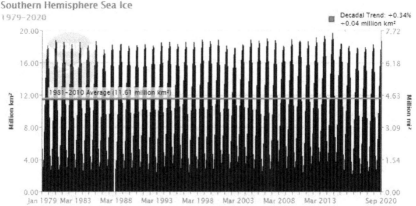

Figure 14-12. Measurements of snow cover and sea ice should be a good indication of climate change. It is encouraging to see that such changes are not large. (NOAA)

SUMMARY

Various types of vortices form in the atmosphere. Some of these may be considered to be unusual because they are not as well known or understood as other more common atmospheric vortices. Dust devils are surface generated vortices that form on hot summer days particularly in desert regions. In mountainous terrain, the mountainado is a vortex that occasionally becomes strong enough to damage houses and other property. The mountainado forms as the air flowing over a topographic barrier interacts with a vortex to strengthen it.

Thermal vortices associated with forest fires and with oil burners designed specially to generate them are of considerable interest. It is not uncommon for a double vortex structure to be generated within thermal plumes. Thermal vortices frequently accompany volcanic eruptions.

Many atmospheric storms undergo metamorphosis. When this happens one of the individual atmospheric storms with its own unique structure is changed into a different type of atmospheric storm. Thus, hurricanes change into frontal cyclones, and dust devils may change into mountainadoes or tornadoes under the right conditions.

Unusual weather patterns may be responsible for very different weather in various parts of the world. Large meanders in the jet stream may provide Alaska with warmer than normal temperatures at the same time that Florida is experiencing below freezing temperatures.

The weather may be unusual because of persistent weather patterns, unusual intensity, or types of storms. The intensity of some hurricanes and tornadoes is such that they must be considered unusual because of this. For example, Hurricane Camille was one of the strongest hurricanes ever to strike the United States, similarly the large tornadoes such as those that struck Lubbock, Topeka, and Wichita Falls were unusual because of their size and intensity.

Various weather elements and surface features may combine to create unusual weather occurrences. Thus, the haboob is caused by the gust front of a thunderstorm which picks up dust from desert areas and generates a dust storm with very low visibility. Many other types of unusual weather are possible. The red rains of France and the blue rains of Greece are caused by dust particles which serve as condensation nuclei within thunderstorms. The rain then takes on the color of the particles that serve as nuclei for the raindrops.

The climate is controlled by a large number of interrelated factors. If one of these, such as temperature is changed, then others are affected as well, such as cloud cover. The amount of snow and ice each year should be a good indicator of climate change. Satellite measurements are available since 1966 and these show very little changes with time.

15

Human Response to Weather

Now that the season's here again; it seems worse every year,
When every advancing squall line fills everyone with fear.
The weather station radar shows a certain kind of blotch,
And the whole surrounding area goes on tornado watch.
One hundred miles on either side of a line from here to there
Marks out the ill-starred region to be ravaged from the air.
The picture tubes and radios spew forth the voice of doom;
Uneasy folks peer nervously through the deepening gloom.
Children sense their parents' fear, begin to cry and fret;
Each ominous announcement reamplifies the threat.
To a terror-stricken observer with myopic sight endowed,
A trailing wisp of vapor looks just like a funnel cloud.
He quickly calls the station, relating what he has seen,
And the "citizen observer" has brought panic to the scene!
Folks dash to the basement, or cower 'neath the bed
As the word "tornado warning" fills anxious hearts with dread.
Our storm watch broadcast network saves many lives, 'tis said,
But the frightened die a thousand deaths, before they're really dead.
Of the thousand ways of dying, if you analyze them all,
The chance of being blown to death is ridiculously small.
So, when the voice assails me, "a fearful storm is brewing,"
I remember the statistics, and continue what I'm doing.
If the awesome funnel gets me, as in the end it might,
I'll have lived my life serenely, not quivering in fright.

-Charles Lacey, Lawrence, Kansas

A very common response of many people to severe and unusual weather is to ignore it and assume that it will never affect them directly. This assumption may cause a person to exhibit such unconcerned behavior as to continue taking photographs of a tornado, prior to, and after the tornado literally blows him off his feet, as happened during the Wichita Falls Tornado. It can also cause a person to spend his Thanksgiving vacation in a community building in a strange city while a blizzard rages outside as happened to hundreds of people in 1979 as a blizzard produced drifts several yards high to close highways and strand thousands of people.

BLIZZARDS

Many people do not have an appropriate respect for the potentially lethal power of the blizzard. Thus, the blizzard is underestimated in contrast to the tornado that is frequently overestimated in terms of its lethal ability. In fact, both of these types of storms cause about the same number of deaths each year. A reasonable response to the threat of blizzards is to recognize their capability of closing major highways, and making travel on streets impossible. Thus, one who insists on traveling through a blizzard as it moves across the United States must realize that it has the capacity for reducing the heated interior of his automobile to outside temperatures by stalling the car where it will run out of gas in a few hours if it is used to keep the heater going.

On November 21, 1979, a blizzard cyclone was dumping over 2 ft. of snow in Cheyenne, Wyoming and Denver, Colorado, before starting to progress eastward. It continued to create blizzard conditions with low temperatures and heavy snowfall. It was no secret that this winter storm would be moving across the Central Plains at a time when many people would be traveling for their Thanksgiving vacations. The inevitable result was stranded motorists and 125 people killed. National Guardsmen in Wyoming rescued more than 2000 travelers − a stalled caravan 2 mi. long − on Interstate 25 north of Cheyenne. Several had to be rescued by helicopter. Motels and hotels became jammed throughout the Midwest as the blizzard moved through Kansas, Nebraska, Iowa, and Minnesota. Many people spent their Thanksgiving stranded in their cars, in motels, National Guard Armories, or municipal auditoriums. In Sidney, Nebraska, about 60 motorists were treated to a turkey dinner by a local church group. Many others were lucky just to keep warm.

Only a little more than a year earlier on January 26, 1978, another blizzard had moved through the Midwest and killed almost as many people. More than 5,700 motorists were stranded along snow-blocked highways in Ohio alone, as Governor James Rhodes called the storm "a killer blizzard looking for victims." The freezing temperatures were an added difficulty to many other Ohio residents as more than 150,000 homes were without electricity for heating for most of the day on January 26th. A federal state of emergency was declared by President Carter and the 5th Army moved in to help stranded motorists and exhausted utility repairmen.

Such examples occur more frequently than most people realize and emphasize the need for proper respect for these winter storms. It is apparent that the proper response is to consider weather forecasts when making plans to travel during winter, and to plan around major winter storms. It is encouraging to note that the abundance of cell phones, four wheel drive automobiles and a more responsive public has lessened the repeat of so many stranded victims in recent years.

TORNADOES

The specific response of a person to tornadoes, in general, is as varied as the number of different types of personalities. Some people who live in cities that have been struck by tornadoes, and other people who are only vaguely aware of tornadoes, are terrified by them. Others consider them a thing of beauty to be included in a work of art or to be photographed as one of nature's most beautiful creations. Most people are more afraid of tornadoes than is warranted by statistics on deaths and injuries caused by them. This, no doubt, is because of the very violent nature of tornadoes.

Many factual stories are circulated after a tornado that may reinforce or initiate fear of these storms. Less publicized, however, is the fact that a typical severe tornado can destroy one thousand houses, most of which have occupants in them, and kill only twenty people. If the destroyed houses contained an average of 3 persons per house, then only 20 people out of the 3000 potential victims would be killed. Based on this example, your chances of being killed by a tornado are less than 1% even if your house is totally destroyed by a tornado.

A particular person's chances for survival during a direct hit by a tornado is certainly improved by applying the best information on safest locations in houses and by properly designing houses and basements for shelter from storms. Statistics show that these responses combined with the increasing accuracy of warnings and forecast information are resulting in fewer deaths each year from tornadoes even though property damage from tornadoes continues to increase.

LIGHTNING

Many people's response to lightning consists of placing a lightning rod on their house or office building as protection from a lightning strike or obtaining insurance for personal property. Many people do not realize that lightning is responsible for as many or more deaths each year than tornadoes; since these deaths are usually single events, they do not make the headlines as often as many other storm related deaths. Individual responses to lightning are probably more rational than reactions to many of the other potentially fatal forms of severe weather. This is probably because the threat from lightning can be recognized by the loud noises and visual displays that accompany thunderstorms. Therefore, if people are aware of the most effective precautions during lightning displays, they are more likely to put them into practice.

HAILSTORMS

Major human responses to hailstorms have consisted primarily of insuring personal property and agricultural crops. The threat of damage from hailstorms varies with geographical location but in many areas a significant cost of crop production is insurance to pay for hail damages, should they occur. Automobile insurance and house insurance normally provide coverage from hail and wind damage for most of the United States.

Even though the National Hail Experiment concerned with modifying hail producing thunderstorms was terminated because of lack of positive results, weather modification for the purpose of reducing hail damage continued for many years prior to the turn of the century.

Many recent years have had more than 5000 major hailstorms with insurance companies paying more than ten billion dollars in damages.

HURRICANES

Our ability to respond to the threat of hurricanes has been greatly improved in the United States since the age of satellites. Since satellites are able to photograph cloud patterns several times each hour for early identification, their long life allows greater preparation time and a more rational response.

Forecasting the exact location of landfall is still a limitation for specific warnings since every effort is made to prevent unnecessary evacuations of cities. The number of people that are forced to evacuate low-lying coastal areas for a typical strong hurricane is about a quarter of a million people. When this number of people are involved, it is not surprising that a few brave but unenlightened people decide to remain behind for a hurricane party as a large storm approaches.

With today's mobile society, it is estimated that more than 75 percent of the population along coastal areas of the United States are inexperienced with hurricanes. In addition, a person who has experienced a previous hurricane is conditioned by that experience. This is not always helpful since many weak hurricanes strike the United States and give a person riding them out a wrong impression of the possible intensity of future storms. Therefore, it is important to realize that hurricanes come with a variety of intensities. When a strong hurricane approaches land, the only rational behavior is evacuation of low-lying areas when advised to do so.

Methods of evacuation and evacuation routes, therefore, become important. Methods of evacuation are a major limitation in many less developed and overpopulated countries. In the United States, however, the automobile offers a rapid means of escaping the fury of these storms. Evacuation procedures and routes have been prepared for many counties along the coast to prevent problems that can develop from short warnings and overcrowded highways. Individuals may also seek to protect property by boarding up windows, or by purchasing metal hurricane protectors that are available commercially to help cover buildings during a hurricane.

DROUGHT

Response strategies to drought are more difficult to plan than for some of the other forms of severe weather. Irrigation systems are frequently used to eliminate drought. However, such solutions require a source of water and also considerable lead time for putting the system into operation.

In the United States, government assistance is frequently provided to drought-stricken areas, just as it is for victims of other forms of severe and unusual weather.

Nations may respond to the threat of drought by the storage of grains and other food products for use in the event of an extended national drought. City governments may react to a drought problem by first requiring the conservation of water, followed by rationing of the water supply. Continued drought may fan the flames of proposals such as piping water in, desalination of sea water, or dragging down an iceberg.

Within the agricultural community, those affected by drought may decide to alter their cropping strategies or seek a temporary job off the farm. Ranchers may liquidate their herds to be replenished later, probably for higher prices, as their pastures are burned into the ground. More drought resistant crops or crops that mature faster may be additional options. Many farm ponds were dug after the dry years of the 1930s. These served as an important source of water during droughts in the 50s and also during more recent droughts.

FLOODS

A major human response to floods, as in other types of severe weather, has been to ignore the potential threat. A significant part of many cities is located within the floodplain of a large river. Housing developments and industries frequently expand into low-lying floodplains. Such planning and construction completely ignore the fact that the floodplain surrounding a river is flat because it is occasionally inundated by water. The return period for floods may be several times per decade or only once per century. As a community with developments in a floodplain realizes the potential threat, levees may be built along the river banks. As these continue to require additional height they may eventually reach the stage where the bottom of the river channel is actually above the lowlands outside the levee area. This makes the threat of flooding even more pronounced.

Dikes may also extend out into the current of a river, such as the Missouri River, to prevent it from eroding the banks. The Mississippi River Commission, established in 1879, has constructed an entire system of levees along the banks of the river. These now total more than 2000 mi. in length and are more than 30 ft. high in some places.

Response to the threat of floods on a national level has also included the establishment of flashflood warning procedures with personnel to operate river forecast offices and the establishing of automated flash flood warning systems. An additional Federal response is the construction of dams to form lakes with the intent of flood control of particular rivers. By controlling the flow of water over a dam, flood losses may be reduced or prevented downstream. If the flow of water over the dam is of sufficient quantity and height to produce electricity this is an additional benefit.

Individuals may respond to the threat of flood by recognizing that floodplains suffer from the threat of floods. Thus, a floodplain would not be selected for a housing development if another choice were available.

Proper reaction to the threat of flashfloods is also very important. A tabulation of the 25 casualties of the Kansas City Plaza flash flood of September 12 and 13, 1977, indicated that 17 of the victims were driving or passengers in cars, 2 were viewing the flood waters, 2 were walking, 1 was electrocuted, and 1 died of a heart attack. The activities of the other two victims prior to the flood were unknown. Many people did not take the rising flood waters seriously. As the water reached the doors of restaurants, the doors were closed. As the flood waters broke windows, the occupants retreated to a higher floor.

Drivers of automobiles are generally observed to minimize the dangers of driving during flash flood warnings. Many instances have been reported, where motorists proceeded through flooded streets when they could see cars stalled ahead. Thus, it is important to emphasize the importance of proper reactions when driving during flash flood warnings.

WEATHERING

Weathering is a term that indicates how people cope with the weather. Several different weather factors may affect people, including temperature, pressure, humidity, air pollutants and air ionization. According to surveys, about 50% of the population is somewhat sensitive and a smaller percentage is very sensitive to the weather. The geographical location where people live affects their sensitivity. In San Diego, for example, the weather does not change very much and people become acclimated to these small changes. If they move to the Midwest, for example, where the weather is more variable, they are more sensitive to weather changes than others who have lived there longer. Sensitivity to weather also changes with age. Babies and people more than 60 years of age are more sensitive than others.

TEMPERATURE

Temperature is one of the primary elements of the atmosphere that affects people. Our bodies try to maintain a constant temperature by such physiological changes as the expansion or contraction of blood vessels. They expand in hot weather to circulate more blood close to the surface of the skin where sweat serves as a cooling mechanism. In very cold weather, the blood vessels contract, and send the blood to deeper parts of the body and this maintains a constant internal body temperature. Our bodies also shiver when we are cold. This is an involuntary response mechanism that helps us to keep warmer.

Heat affects many people emotionally. An increase in violent crimes has been observed in hot weather. Warmer temperatures also tend to dull the intellect. Records were kept for many years on the test scores on civil service exams which showed that more people failed the exam in the months of July and August than any other month.

HUMIDITY

Humidity exerts an effect on people, but it is generally less than the effect of temperature. The combination of high humidity and warm temperature leads to uncomfortable conditions for most people. Less water evaporates from the skin, under these conditions and this is the bodies primary cooling mechanism, since heat from the skin is used in evaporating water.

Humidity also affects the skin directly, since it contracts with low humidity and expands with high humidity. Scar tissue is not as flexible as ordinary skin tissue. When the air is very dry, ordinary skin contracts more than the scar tissue and exerts sufficient pressure to cause severe pain to some people.

Another effect of humidity is exerted through the formation of fog, low visibility and clouds. Some people become depressed under these conditions.

One study showed that when the atmosphere was dry, people checked out more serious books from the library and when it was moist, they checked out books for lighter reading. Therefore, people's moods seem to swing with the humidity, cloudiness or their perceived humidity conditions.

PRESSURE

Atmospheric pressure affects parts of the body that are sealed from rapid pressure changes. These include such joints as knees, elbows and shoulders. The brain is also sealed from rapid weather changes and some people have migraine headaches as the pressure changes, while others have aches in their joints.

Severe sunburn, which is more likely to occur with high pressure, causes other effects on a person in addition to the altered skin. More hormones are produced by the pituitary gland and these affect the digestive tract and blood pressure.

A very simple traffic display at a state fair resulted in an interesting conclusion. The exhibit was designed to test a person's reaction time by measuring the length of time, after a light changed from green to red, required for a person to push a button. After a person's reaction time was tested it was displaced for viewing. The number of persons who took the test and their reaction time was recorded. When the results were compared for each day after more than 20,000 people had taken the test it was discovered that the average reaction time was very different each day. Since it was unlikely that the differences could be explained by variations in groups of people each day, other explanations were sought. When the results were compared with the weather on the days when the tests were taken it was found that the fast reaction times corresponded to high-pressure conditions. The slowest reactions occurred as a low-pressure system was approaching and during the time while it passed. The differences in reaction times were so different that they were valid statistically and no other explanations seemed plausible.

Perhaps the automobile accidents that are more numerous under low- pressure conditions that are normally attributed to cloudy skies, lack of visibility, or wet pavement, are also affected even more by a person's reaction time. A person becomes accustomed to a particular automobile and soon learns to stop at an exact distance. If the distance required for stopping under a low-pressure system is only a few feet greater than with a high-pressure system, this may be sufficient to cause numerous accidents. Therefore, it is worth noting and remembering that greater distance is required for stopping a car when the atmospheric pressure is low, because the time is longer between the moment a person perceives an emergency and the time the car actually stops.

PSYCHOMETEOROLOGY

A number of symptoms have been identified which are commonly associated with the weather. People who are sensitive to the weather as well as those who are less sensitive report the symptoms. Some of these are tiredness, bad moods, disinclination to work, head pressure, restless sleep, headaches, impaired concentration, difficulty in falling asleep, nervousness, bone fracture pains, and visual flickering.

Psychometeorology deals with the interaction of moods and weather. When people were asked about their general state of health, as they perceived it, they tended to use such terms as "very tired" when the weather was warm and moist and "feel bad" when a change in the weather was coming. They stated that their desire to work declined or their work productivity was lessened by extremely good weather. Such psycometeorological effects are much harder to substantiate than more direct effects, such as the additional suffering from rheumatoid arthritis during cold and moist weather.

MOOD AND WEATHER

Observations have shown that psychiatric patients were more irritable and in gloomier moods when the skies were cloudy. On the other hand, high pressure systems with sunny skies may produce the opposite effect. One study showed that sunlight suppresses melatonin which is a hormone that tends to produce a depressed state. If sunlight interferes with its production then this represents a direct relationship between the condition of the atmosphere and a person's mood.

Weather changes affect how people relate to each other. Benjamin Franklin, many years ago, advised, "Do business with men when the wind is in the west, when the barometer is high." This remains very good advice, and was verified by a recent study conducted by stopping strangers on the street and asking them a series of questions. The number of questions they were willing to answer before they turned away provided striking results. Americans were much more willing to answer questions on a sunny than on a cloudy day. This difference was attributed to people's moods which were related to the weather.

Some people sink into a more depressed state in the winter months and apparently don't really notice it until spring comes and other people are in a better mood. This is one explanation for the fact that more suicides occur in spring.

Evidence indicates that a person's resistance to viruses is related to mood and mental state and these may be related to weather or other causes. Depressed persons are more susceptible to virus attacks.

RESPONSE TO POSITIVE IONS

Evidence is accumulating that physiological effects can be produced in many people by severe thunderstorms. The positive ion concentration near the ground in a typical thunderstorm is 1,000 to 2,000 ions per cubic centimeter, but varies considerable as lightning discharges occur. Although some people are more sensitive than others, an inhalation of positive ions may increase the level of serotonin in the blood. Serotonin is a powerful hormone that acts in the midbrain on the sleep process, transmission of nerve impulses, and the development of mood.

One investigator, F.G. Sulman, found that large concentrations of positive ions are associated with hot, dry desert winds. They are also associated with chinook winds.

At least three major effects of high concentrations of positive ions have been identified. The serotonin irritation syndrome occurs when an excess of serotonin is produced by the body. It is characterized by headaches, irritability, sleeplessness, and heart pain.

A second response to large positive ion concentrations is the exhaustion syndrome. In the exhaustion syndrome, the body responds first with an euphoric mood as a result of positive ions triggering the body's release of adrenaline. The adrenaline causes a burst of energy, but this is soon followed by exhaustion. Some unfortunate individuals become unable to function normally as additional weather-related symptoms take over.

The third effect is the hyperthyroid response. This occurs when a person's thyroid gland is stimulated and produces excess thyroid hormone. The symptoms are similar to the exhaustion syndrome with an initial burst of energy followed by exhaustion and other reactions such as headache, heart pain or emotional instability.

It is important for people who are extremely weather-sensitive to be aware of the physiological changes that a severe thunderstorm, chinook or desert wind can produce so that their reactions can be anticipated. Counteracting, negative-ion generating machines are available commercially, but it is questionable whether some of them have the capacity to cancel the positive ions in a very large room.

WEATHER SENSITIVITY

Different types of people have been identified in an investigation by the American Bioclimate Institute according to the effect of weather on them. People were grouped according to "cold-front people" and "warm-front people". The cold-front person is typically tall, thin and has long limbs. They find cold air much more unpleasant than other people. The warm-front people are generally short, stocky, and have short limbs. They find warm air very uncomfortable.

Most people are not very sensitive to individual variations in responses to weather or to single atmospheric elements such as temperature but some are very sensitive. Not realizing that differences in weather sensitivity do exist, may affect our interpersonal relationships. Therefore, it is important to know that we don't all react the same way to the weather around us, or even to single atmospheric features such as pressure or temperature. For example, the thermostat setting varies greatly in peoples houses. Older people generally set it higher in winter than younger people. Further, if a cold-front person lives with a warm-front person they may have real differences in the choice of thermostat setting because they are affected differently by their temperature environment.

A test has been devised by Rosen in a book, *Weathering,* to reveal your sensitivity to the weather. It allows a greater range in response to the weather than the two categories of cold-front people or warm-front people. Your own breakdown into one of six categories is determined by summing the numbers assigned to the following characteristics listed on the next page.

SELF TEST FOR WEATHER SENSITIVITY

Category 1 is physique characteristics.

- If you are lean, slender or lanky, add 3 points.
- If you are broad, stocky and stout, add 1 point.

Category 2 is temperament characteristics. You gain points from as many questions as apply to you.

- If you tend to me extroverted and jolly, add 1 point.
- If you are often emotionally changeable, add 3 points.
- If you tend to be easily led or acquiescent, add 3 points.
- If you are often irritable or moody, add 1 point.
- If you tend to be depressed or pessimistic, add 2 points.
- If you are often shy, inhibited or private, add 3 points.
- If you tend to be nervous, add 4 points.

Category 3 is socioeconomic status.

- If you are professional, or upper class, add 3 points.
- If you are blue collar, clerical or laborer, add 3 points

Category 4 is your age.

- If you are 10 to 19 years old, add 3 points.
- If you are 20 to 29 years old, add 2 points.
- If you are 30-39 years old, add 1 point.
- If you are 40-49 years old, add 2 points.
- If you are 50-59 years old, add 3 points.
- If you are older than 59, add 4 points.

Category 5 is gender.

- If you are female, give yourself 3 points.

NOW ADD UP YOUR TOTAL POINTS.

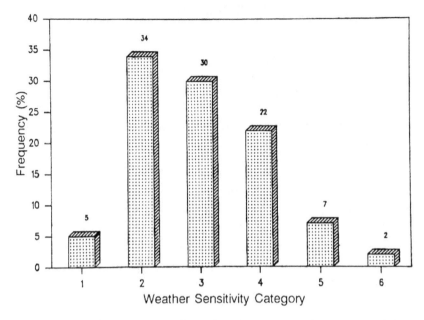

Figure 19-1 The percentage of people who were weather resistant (1), weather receptive (2), weather sympathetic (3), weather susceptible (4), weather responsive (5) or scored in the weather dominated category (6).

Your weather sensitivity is determined by your total score from all five categories. If you scored, 0-5, you are **weather resistant**. Generally, you are indifferent to weather changes. If you scored 6-10, you are **weather receptive**; you are aware of your reactions to weather changes. A score of 11-15 indicates that you are **weather sympathetic**; you are never indifferent to weather changes. If you scored 16-20, you are **weather susceptible**; you are always in touch with symptoms that the weather induces. Some people are **weather responsive** (score of 21-25). If you are in this category every weather change is felt in your body. A few people score greater than 25 (**weather dominated**), where severe pain or pleasure accompanies weather changes.

This test was given to several hundred college students taking my Unusual Weather class to determine a typical distribution for the various categories (Figure 19-1). The results showed that 5% of the people were weather resistant while 34% were weather receptive, and 30% were weather sympathetic. Many people were weather susceptible (22%) and 7% were weather responsive. Only 2% of them were in the weather dominated category.

Interviews with individuals who scored in the extreme categories were conducted to provide a check on the results. Those individuals who were rated "weather resistant" paid very little attention to the weather, as the test predicted, while those rated "weather dominated" felt that they, indeed, were dominated by the particular weather they were experiencing.

CONCLUSIONS

In general, human response to weather varies with the individual and with the type of weather. In both categories the range of voluntary responses is from intelligent, informed responses to unprepared, uninformed responses. It is important to note that voluntary responses can be changed. The person who is caught in a blizzard once is more likely to make better decisions during the threat of the next winter storm. Furthermore, with the proper understanding it is not necessary to experience all aspects of severe and unusual weather in order to respond appropriately.

Involuntary responses to weather are also very different for individuals and range from weather resistant to weather domination. Active weather factors include positive ions from severe thunderstorms or chinooks, as well as variations of ordinary weather elements.

It may be helpful to understand how weather can affect us in ways beyond our immediate control. Perhaps the information contained in this book will play an important part in helping you make better plans and decisions as you better understand, and interact with, various types of normal, severe and unusual weather in the future.

Joe R. Eagleman

INDEX

Air mass, 11, 13-17, 27, 57
Anticyclonic vortex, 73, 76, 78, 84, 178, 251
Anticyclonic tornado, 251, 252
Apparent temperature, 48

Basements, 1221-124, 131, 137
Big Thompson Flood, 228, 229
Blizzard, 28, 45, 260, 261
Bow echo, 59, 248

Chinook, 28, 45-48, 253, 269, 270
Climate change, 254, 256, 258
Conservation of Angular Momentum, 6-8, 27, 68, 89, 214
Cyclogenesis, 20-23, 27

Doppler radar, 71, 73, 74, 78, 100, 106, 108, 117
Double vortex, 71-75, 79-88, 100, 108, 173-178, 246
Dry Adiabatic Lapse Rate, 51, 52
Dry Line, 54, 55, 57-59, 63, 64, 67

Derecho, 248
Double vortex, 74-78, 100, 173, 174, 176-178
Dust Devil, 106, 187, 243-246, 250, 251, 257
Dynamic updraft, 77, 88, 106, 107

Easterly waves, 183
Eddy tornado, 251
El Nino, 248, 249
Evapotranspiration, 236-238, 242

Flash flood warning, 232, 265
Franklin, Benjamin, 140-142, 160, 268
Freezing rain, 33-36, 41, 37
Frostbite, 43

Glaze, 45, 48
Gust front, 69, 76, 77, 248, 253, 257

Haboob, 252, 253, 257

Hail, 168-175, 178, 179, 180, 262
Hailstorm, 1, 162-166, 170-180, 262
Hook echo, 58, 64, 73, 99, 100, 101, 104
Hurricane category, 93, 187, 198, 199, 205, 209, 212, 213, 217
Hurricane formation, 182, 186, 235
Hurricane names, 188, 189

Ice Pellets, 34, 35, 47, 150, 151, 160, 254
Ice storm, 42, 48
Intertropical convergence zone, 185, 196, 235, 236
Inversion, 34, 35, 53, 56, 64, 67, 97, 185
Inverted barometer, 116, 193
Isobars, 184

Jetstream speed, 11, 176-179

La Nina, 248, 249
Laboratory tornado, 107-116
Latent heat, 9, 42, 51, 89, 185, 194, 196, 209, 214
Lifted index, 64, 66
Lightning
   Ball, 142, 146, 147, 160
   Bolt from the blue, 145, 146, 160
   Forked, 142, 143, 160
   Rods, 154, 161
   Sheet, 144, 145
Longwave cyclone, 27-29, 239
Low level jet, 54

Mesocyclone, 76, 88, 99, 100-107, 177
Metamorphosis, 250, 251, 257
Microburst, 77. 89
Mississippi river flood, 224
Model houses, 133, 191
Moist adiabatic lapse rate, 51
Moisture tongue, 64, 66

Mountainado, 251, 257

National Hurricane Center, 210, 217
National Weather Service, 42, 82, 101, 166, 215, 241
Northeaster, 33, 38

Ocean swells, 193

Rapid City flood, 230
Return streamer, 152, 153, 160
Roof damage, 93, 191

St. Elmo's fire, 146, 147, 160, 254
Santa Anna wind, 253, 254
Snow flurries, 35
Soil moisture, 236, 238
Squall line, 57-59, 67, 87, 99, 171
Superbolt, 145

Thermal vortices, 247, 257
Thermoelectric effect, 149
Thunder, 148
Thunderstorm structure, 74-82
Thunderstorm splitting, 80, 81
Tornado appearance, 94, 95
Tornado damage, 122-135
Trough cyclone, 31, 33

Upper air divergence, 20-22, 61, 62
Urban heat island, 82, 256

Vortex twinning, 243-245
Vorticity, 21-23, 27, 82, 57, 67, 76

Warning
   Blizzard, 29, 37
   Flash flood, 232, 265
   Heavy snow, 35
   Hurricane, 215-217
   Tornado, 64, 65, 99, 101-104, 216, 259
Waterspout, 89, 109

Weak echo vault, 174, 175
Wind chill, 48
Wing shear, 21-23, 27, 54, 60, 61, 75, 113, 170, 245
Winter travel, 29, 44

ABOUT THE AUTHOR

Joe R. Eagleman (1936-) was born on a farm near West Plains Missouri. He received the PhD from the University of Missouri in 1963 and served as professor at the University of Kansas for 39 years and continues as Professor Emeritus. He taught thousands of students about Atmospheric Science through his courses there and many thousands more through four different textbooks used by over a hundred universities over a span of several decades.

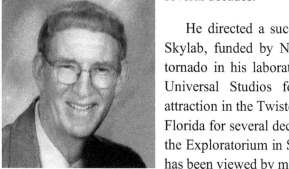

He directed a successful experiment on Skylab, funded by NASA, and invented a tornado in his laboratory that was used by Universal Studios for a 50 ft. tornado attraction in the Twister Building in Orlando Florida for several decades. It can be seen at the Exploratorium in San Francisco where it has been viewed by millions of people. He is also the author of a technical book on severe thunderstorms that includes his tornado safety research which resulted in changes that were adopted nationally.

His autobiography, *Name Your Price*, tells of his early life on a farm where he was the 11th of 12 children. It includes his work as a scientist as well as a number of unusual hobbies including those as an artist, musician, luthier, marksman, taxidermist, world traveler and other endeavors.

He is the author of books on making musical instruments and has recorded five albums of his original music with a guitar that he made. He has a gallery of paintings on Fine Art America.

For more information see http://www.JoeEagleman.com.

Printed in the USA
CPSIA information can be obtained
at www.ICGtesting.com
LVHW011241141223
766422LV00004B/461